Data Analytics for Intelligent Systems

Techniques and solutions

Online at: https://doi.org/10.1088/978-0-7503-5417-2

IOP Series in Next Generation Computing

Series editors

Prateek Agrawal

University of Klagenfurt, Austria and Lovely Professional University, India

Anand Sharma

Mody University of Science and Technology, India

Vishu Madaan

Lovely Professional University, India

About the series

The motivation for this series is to develop a trusted library on advanced computational methods, technologies, and their applications.

This series focuses on the latest developments in next generation computing, and in particular on the synergy between computer science and other disciplines. Books in the series will explore new developments in various disciplines that are relevant for computational perspective including foundations, systems, innovative applications, and other research contributions related to the overall design of computational tools, models, and algorithms that are relevant for the respective domain. It encompasses research and development in artificial intelligence, machine learning, block chain technology, quantum cryptography, quantum computing, nanoscience, bioscience-based sensors, IoT applications, nature inspired algorithms, computer vision, bioinformatics, etc. and their applications in the areas of science, engineering, business, and the social sciences. It covers a broad spectrum of applications in the community, including those in industry, government, and academia.

The aim of the series is to provide an opportunity for prospective researchers and? experts to publish works based on next generation computing and its diverse applications. It also provides a data-sharing platform that will bring together international researchers, professionals, and academics. This series brings together thought leaders, researchers, industry practitioners, and potential users of different disciplines to develop new trends and opportunities, exchange ideas and practices related to advanced computational methods, and promote interdisciplinary knowledge.

A full list of titles published in this series can be found here: https://iopscience.iop.org/bookListInfo/iop-series-in-next-generation-computing.

Data Analytics for Intelligent Systems

Techniques and solutions

Edited by

Sachin Taran

Delhi Technological University, Delhi, India

Chhavi Dhiman

Delhi Technological University, Delhi, India

Manjeet Kumar

Delhi Technological University, Delhi, India

IOP Publishing, Bristol, UK

Permission to make use of IOP Publishing content other than as set out above may be sought at permissions@ioppublishing.org.

Sachin Taran, Chhavi Dhiman and Manjeet Kumar have asserted their right to be identified as the editors of this work in accordance with sections 77 and 78 of the Copyright, Designs and Patents Act 1988.

ISBN   978-0-7503-5417-2 (ebook)
ISBN   978-0-7503-5415-8 (print)
ISBN   978-0-7503-5418-9 (myPrint)
ISBN   978-0-7503-5416-5 (mobi)

DOI   10.1088/978-0-7503-5417-2

Version: 20240201

IOP ebooks

British Library Cataloguing-in-Publication Data: A catalogue record for this book is available from the British Library.

Published by IOP Publishing, wholly owned by The Institute of Physics, London

IOP Publishing, No.2 The Distillery, Glassfields, Avon Street, Bristol, BS2 0GR, UK

US Office: IOP Publishing, Inc., 190 North Independence Mall West, Suite 601, Philadelphia, PA 19106, USA

Contents

6 Imbalanced class problem analysis for lung cancer detection using convolutional neural networks

Om Mishra, Deepak Parashar, Amit Kukker, Aditi Rao, Ananya Srivastava, Anahita, Rajan Mishra and P S Kavimandan

12 Misleading multimodal news dataset for detecting fraudulent content 12-1

Deepika Varshney, Mohit Aggarwal and Neetu Saradana

13 Structural crack detection, segmentation, and classification: a review 13-1

*Basavaraj Katageri, Rajashri Khanai, Rajkumar V Raikar,
Dattaprasad A Torse and Krishna Pai*

14 A systematic review of fault detection in hardware and software systems 14-1

Mayank Yadav and Ruchika Malhotra

15 ACPSOD-Net: a deep atrous convolution pooling based network for salient object detection   15-1

Bhagyashree V Lad, Mohammad Farukh Hashmi and Avinash G Keskar

Preface

In today's rapidly evolving world, data are being generated at an unprecedented pace. IOT devices, machine-generated logs, and sensor readings generate vast amounts of data every second. These data contain valuable information and insights that can drive intelligent decision-making systems.

This book serves as a comprehensive guide to the principles and techniques of data analytics. Throughout the book, we emphasize the practical aspects of data analytics by providing real-world examples and case studies from different domains. The book provides an intelligent framework for data analytics for modelling concepts and applications using different data dimensions. The applications considered are biomedical engineering, fault detection in machinery, object segmentation and classification, image enhancement under extreme environmental conditions, recommender systems, and image fusion.

Each chapter begins with problem formulation and explains the significance of developing the framework in a certain way, then delves into various statistical and machine learning techniques that enable predictive modelling, clustering, classification, and anomaly detection. The chapters can be read independently by researchers, R&D engineers, and undergraduate students who wish to explore biomedical signal processing, computer vision, fault and crack detection, and recommender systems research.

Chapter 1 presents a precise ECG QRS complex detector using a whale optimization algorithm (WOA) optimized fractional order digital differentiator. Chapter 2 explores focal and non-focal EEG signal classification using the Wigner–Ville distribution and deep feature extraction. Chapter 3 describes multi-channel EEG-based affective emotion identification using a dual-stage filtering approach. Chapter 4 presents variational mode of decomposition-based entropy features for classifying myopathy, neuropathy, and normal EMG signals. Chapter 5 employs a wavelet transform followed by support vector machine for epileptic EEG signal classification. Chapter 6 addresses imbalanced class problems for lung cancer detection using convolutional neural networks. Chapter 7 demonstrates an end-to-end content aware generative adversarial network based method for multimodal medical image fusion. Chapter 8 describes infrared thermography in diagnosing macular edema. Chapter 9 presents variants of generative adversarial networks for underwater image enhancement. Knowledge graphs are explored in chapter 10 for job tree analysis and recommendation using ensemble models. Chapter 11 presents variants of the SVD algorithm for recommender system implementation. Chapter 12 introduces a misleading multimodal news dataset for detecting fraudulent content. Chapter 13 presents a review of structural crack detection, segmentation, and classification. Chapter 14 presents a systematic review of fault detection in hardware and software systems. Chapter 15 describes ACPSOD-Net: a deep atrous convolution pooling based network for salient object detection.

We hope that *Data Analytics for Intelligent Systems: Techniques and Solutions* becomes a valuable resource for anyone seeking to harness the power of data analytics to drive intelligent decision-making. Whether you are a seasoned practitioner or a newcomer to the field, the insights and techniques presented in this book will empower you to extract actionable knowledge from data and unlock the full potential of intelligent systems.

Acknowledgements

Dr Taran expresses his heartfelt indebtedness to his parents, wife, and lovely daughters Vedika and Manogyaa for their unending inspiration throughout the completion of this book.

Dr Dhiman expresses his hearty gratitude to the Almighty for His blessings, her dear husband, son Arya, parents, and all respected teachers for their invaluable contribution and consistent guidance and support.

Dr Manjeet Kumar expresses his hearty gratitude to his family members, wife Pratibha, son Nuvesh, and parents, for rendering unfailing support during the editing of the work, the unsung heroes in our lives, and all respected teachers for their consistent guidance.

We sincerely thank our University Vice-Chancellor, Head of the Department, and colleagues in the Department of Electronics and Communication Engineering at Delhi Technological University. We want to thank all our friends, well-wishers, and all those who keep us motivated to do more and more, better and better. We sincerely thank all the contributors for their writing on the relevant theoretical background and real-time applications of data analytics for intelligent systems. We are also deeply grateful to many whose names are not mentioned here but whose help during this work we appreciate and wish to acknowledge.

We humbly thank Dr John Navas, Senior Commissioning Manager, and all the editorial staff (IPEM–IOP) of IOP Publishing for their great support, necessary help, appreciation, and quick responses. We also thank IOP Publishing for allowing us to contribute to a relevant topic with a reputed publisher. We are grateful to all those with whom we have enjoyed working during this project.

This book is dedicated to all researchers and innovators pursuing their research journey for continuous advancement. Finally, we would also like to thank God for showering us with His blessings and strength to perform this type of novel and quality work.

Editor biographies

Sachin Taran

Sachin Taran is a motivated teaching professional with approximately eleven years of teaching and research experience in electronics and communication engineering. Dr Taran is working currently as an Assistant Professor at Delhi Technological University (DTU), New Delhi. He received his PhD degree from the Indian Institute of Information Technology, Design and Manufacturing, Jabalpur, India, and completed a postdoc at Nanyang Technological University (NTU), Singapore. Dr Taran served as an Assistant Professor at the Department of Electronics and Communication at Shangvi Innovative Academy, Indore, India, from 2009–2010. He served as an Assistant Professor at the Department of Electronics and Communication at Medicaps University, Indore, India, from 2010–2015. He is a member of various profession societies such as IETE, IEEE, ISTE, and IAENG. He has organized several expert talks, workshops, faculty development programs, and chaired technical sessions for different international conferences.

Chhavi Dhiman

Chhavi Dhiman received her BTech from Indira Gandhi Delhi Technological University for Women (IGDTUW) (formerly known as IGIT, GGSIPU), Delhi, India, in 2011, and her MTech and PhD from Delhi Technological University (DTU) Delhi, India, in 2014 and 2019, respectively. She is currently working as an Assistant Professor in the Department of Electronics and Communication Engineering, Delhi Technological University. Her research interests include computer vision, deep learning, pattern recognition, image processing, and human action and activity recognition, visual captioning, face anti-spoofing, and pedestrian intention prediction in videos. She has been awarded the Commendable Research Award, and Premium Research Award in the years 2020 and 2021, respectively, by Delhi Technological University, Delhi, India. She is a reviewer for various transactions and journals of IEEE, IET, and Elsevier. She has served as a Subject Matter Expert at TCSions since 2021. Her total citations are 564 with an h-index of 9.

Manjeet Kumar

Dr. Manjeet Kumar received the B.Tech degree in Electronics and Telecommunication Engineering from Kurukshetra University, Kurukshetra, India in 2008, and the M.Tech degree in Signal Processing from Guru Gobind Singh Indraprastha University, Delhi, India, in 2011, and the Ph.D. degree in from the Department of Electronics and Communication Engineering, Netaji Subhas Institute of Technology (NSIT), Delhi affiliated to University of Delhi, India, in 2017. He served as Assistant Professor the department of Electronics and Communication Engineering, Bennett University, Greater Noida from June 2016 to July 2020. From July 2020, he has been working as Assistant Professor in the department of Electronics and Communication Engineering at Delhi Technological University, Delhi.

He has authored one book, more than fifty-five research articles and thirty-five conference papers in reputed international journals and conferences. He also served as a reviewer in many International Journals. His research interests include Signal processing, Biomedical signal processing, Image processing, Fractional systems, Optimization algorithms, Nature-inspired algorithms, Artificial Intelligence in Healthcare, Signal analysis using Wavelet Transform, Wavelet filter banks, Adaptive filtering, Linear and nonlinear system identification, Healthcare assistive techniques, and Low-power biomedical circuit design, ECG detection, ECG Classification, PPG Signal Analysis, Heart rate estimation and Blood pressure estimation, Non-Stationary signal analysis, IoMT. He has been awarded with "Premium Research Award" in 2022 and "Commendable Research Award" in 2021 and 2022 by Delhi Technological University, Delhi, India. His total citations are 1527 with h-index 23 and i-10-index 41. He has been included in the prestigious list of the World's Top 2% Scientists prepared by Stanford University for 2021, 2022, and 2023.

List of contributors

Mohit Aggarwal
G L Bajaj Institute of Technology and Management, Greater Noida, India

Ashwini Bakde
Department of Radio-Diagnosis, All India Institute of Medical Sciences, Nagpur, India

Manisha Das
Department of Electronics and Communication Engineering, Visvesvaraya National Institute of Technology, Nagpur, India

Chhavi Dhiman
Delhi Technological University, India

Sanjoli Goyal
Department of Computer Science and Engineering and Information Technology, JIIT, Noida, India

Deep Gupta
Department of Electronics and Communication Engineering, Visvesvaraya National Institute of Technology, Nagpur, India

Mohammad Farukh Hashmi
Department of Electronics and Communication Engineering, National Institute of Technology Warangal, India

Muskan Jain
Department of Computer Science and Engineering and Information Technology, JIIT, Noida, India

Ruchi Jain
Sagar Institute of Science and Technology, India

Kranti S Kamble
ECE, VNIT, Nagpur, India

Vikram Singh Kardam
Delhi Technological University, New Delhi, India

Anahita Karthik
Symbiosis Institute of Technology, Symbiosis International (Deemed University), Pune, India

Basavaraj Katageri
KLE Technological University Dr MSSCET, Belagavi Campus, India

P S Kavimandan
Bharati Vidyapeeth Deemed to be University College of Engineering, Pune, India

P V Keshava Krishna
Delhi Technological University, New Delhi, India

Avinash G Keskar
Department of Electronics and Communication Engineering, Visvesvaraya
National Institute of Technology, India

Rajashri Khanai
KLE Technological University Dr MSSCET, Belagavi Campus, India

Smith K Khare
Aarhus University, Department of Electrical and Computer Engineering,
Finlandsgade 22, 8200 Aarhus N, India

Amit Kukker
Symbiosis Institute of Technology, Symbiosis International (Deemed University),
Pune, India

Manjeet Kumar
Delhi Technological University, India

Bhagyashree V Lad
Department of Electronics and Communication Engineering, Visvesvaraya
National Institute of Technology, Nagpur, Maharashtra, 440010 India

Ruchika Malhotra
Delhi Technological University, India

Seema Mehla
Assistant Professor, Department of Computer Engineering & Applications, GLA
University Mathura

Virender Kumar Mehla
APJ Abdul Kalam Technical University Lucknow, India

Om Mishra
Symbiosis Institute of Technology, Symbiosis International (Deemed University),
Pune, India

Rajan Mishra
Madan Mohan Malviya University of Technology, Gorakhpur, India

Sukumar Nagineni
Indian Institute of Technology, Roorkee, India

Chandan Nayak
VIT-AP University, India

Krishna Pai
KLE Technological University Dr MSSCET, Belagavi Campus, India

Deepak Parashar
Symbiosis Institute of Technology, Symbiosis International (Deemed University), Pune, India

Amit Patil
Department of Computer Science and Engineering and Information Technology, JIIT, Noida, India

J Persiya
Vellore Institute of Technology, Chennai, India

P Prakash
Madras Institute of Technology, Chennai, India

Kemal Polat
Bolu Abant Izzet Baysal University, India

Rajkumar V Raikar
KLE Technological University Dr MSSCET, Belagavi Campus, India

Aditi Rao
Symbiosis Institute of Technology, Symbiosis International (Deemed University), Pune, India

S Mohamed Mansoor Roomi
Thiagarajar College of Engineering, Madurai, India

Neetu Sardana
Department of Computer Science and Engineering and Information Technology, JIIT, Noida, India

A Sasithradevi
Vellore Institute of Technology, Chennai, India

Joydeep Sengupta
ECE, VNIT, Nagpur, India

Ananya Srivastava
Symbiosis Institute of Technology, Symbiosis International (Deemed University), Pune, India

Komal Tahiliani
Sagar Institute of science and Technology, India

Sachin Taran
Delhi Technological University, New Delhi, India

Dattaprasad A Torse
KLE Technological University Dr MSSCET, Belagavi Campus, India

Abhuday Tripathi
Sagar Institute of Science and Technology, India

Deepika Varshney
Department of Computer Science and Engineering and Information Technology, JIIT, Noida, India

M Vijayalakshmi
Vellore Institute of Technology, Chennai, India

Mayank Yadav
Delhi Technological University, India

Contributor biographies

Mohit Aggarwal

 **Mohit Aggarwal**, currently a PhD research scholar at Delhi Technological University, completed a BTech degree from Bharati Vidyapeeth Deemed University, Pune, in 2014 and an MTech degree from Motilal Nehru National Institute of Technology, Allahabad, in 2017. He is currently working as an Assistant Professor at G L Bajaj Institute of Technology and Management, Greater Noida. He has more than five years of experience in teaching and research. He worked for three years as an Assistant Professor under the Government of India TEQIP Project at Madhav Institute of Technology and Science, Gwalior. He has also worked at Galgotias University, Greater Noida for over a year. He has recently received a Research Excellence Award from DTU for publishing a research article in an SCIE indexed journal. He has participated in more than 40 FDPs, training programs, seminars etc. He is a member of the Seismological Society of America.

Ashwini M Bakde

 **Ashwini M Bakde** is currently an Associate Professor with the Department of Radio-Diagnosis, All India Institute of Medical Sciences, Nagpur, India. Her current research interests include breast imaging, cardiothoracic imaging, and neuroimaging. Dr Bakde is a Life Member of the Indian Radiological and Imaging Association (IRIA).

Manisha Das

 **Manisha Das** is currently a research scholar with the Visvesvaraya National Institute of Technology, Nagpur. Her research interests include medical image processing and computer vision.

Sanjoli Goyal

 **Sanjoli Goyal** is a highly motivated final-year student pursuing a Bachelor of Technology degree in information technology at Jaypee Institute of Information Technology. She has a keen interest in data analytics and has gained significant experience in this area through various academic projects.

As part of her academic pursuits, she worked on a research project titled *Leveraging Knowledge Graph for Analysis and Recommendation of Jobs*. The project involved analysing resumes and job descriptions using Neo4j, a popular graph database, to recommend jobs to the resumes.

After graduation, she will joined Optum, UnitedHealth Group as a software engineer. Her academic and research experience in data analytics and graph databases have provided a strong foundation for her future career in this area.

Deep Gupta

Deep Gupta is currently an Assistant Professor with the Electronics and Communication Engineering Department, Visvesvaraya National Institute of Technology, Nagpur, India. His research interests include deep learning, computer vision, medical signal and image processing, ultrasound imaging, and cognitive and behavioural analysis. Dr Gupta has been a Life Member of the Ultrasonic Society of India since 2016, and was a recipient of the Dr T K Saksena Memorial Award given by the Ultrasonic Society of India.

Mohammad Farukh Hashmi

Mohammad Farukh Hashmi received a BE degree in electronics and communication engineering from RGPV Bhopal University, an ME degree in digital techniques and instrumentation from SGSITS Indore/RGPV Bhopal University in 2010, and a PhD degree from VNIT Nagpur under the supervision of Dr A G Keskar. He is currently an Assistant Professor with the Department of Electronics and Communication Engineering, NIT Warangal. His current research interests are computer vision, machine vision, machine learning, embedded systems, digital signal processing, image processing, and digital IC design, etc. Dr Hashmi is a member of IEEE, LMIETE, ISTE, and LMIAENG.

Muskan Jain

Muskan Jain earned a BTech in Information Technology from Jaypee Institute of Information Technology, Noida. During her undergraduate studies, Jain focused on the field of knowledge graphs and social network analysis. Her college projects in her third and fourth years were based on developing knowledge graph-based systems, showcasing her expertise in this area.

Jain's professional journey began immediately after completing her final year when she secured an internship as a backend developer at MakeMyTrip, a renowned travel technology company.

Prior to her academic and professional achievements, Muskan completed her schooling at Delhi Public School in Meerut, where she consistently excelled in academics, earning a gold medal for her outstanding performance for seven consecutive years.

Ruchi Jain

Ruchi Jain completed her BE and MTech degrees in computer science and engineering at RGPV University, Bhopal, Madhya Pradesh in 2010 and 2014, respectively. Her MTech specialization was in network security. She is currently working as an Assistant Professor at Sagar Institute of Science and Technology (SISTec), Bhopal. She is also a technical trainer with a keen interest and expertise in network security, machine learning, and deep learning. She is proficient in programming languages such as Python, Django, React, HTML, and CSS. She has leveraged these skills in her work as an assistant professor and technical trainer, where she has developed web applications for data analytics and visualization. She has published several papers in the field of machine learning and deep learning. She is also a skilled technical trainer and has conducted workshops on programming languages and web development frameworks.

Kranti Kamble

Kranti Kamble received a BTech degree in electronics and telecommunication engineering from Shri Guru Gobind Singhji Institute of Engineering and Technology, Nanded, Maharashtra, India, in 2012, and an MTech degree in electronics and telecommunication engineering from Veer Jijamata Technological Institute, Mumbai, India, in 2015. From 2016–2019 she served as Assistant Professor in Electronics and Telecommunication Engineering, Veer Jijamata Technological Institute, Mumbai, India. She is currently a PhD student at Visvesvaraya National Institute of Technology, Nagpur, India. Her research interests include biomedical signal processing and image processing, time–frequency analysis, deep learning, machine learning, and artificial intelligence.

Vikram Singh Kardam

Vikram Singh Kardam received his MTech degree in signal processing and digital design engineering from Delhi Technological University, Delhi, India, in 2017 and BTech degree in electronics and communication engineering from Chhatrapati Shahuji Maharaj University Kanpur, UP, India, in 2007. He joined Galgotias College of Engineering and Technology, Noida, UP, in 2017 as an Assistant Professor in the Electronics Department. Currently he a PhD student at Delhi Technological University,

Delhi, India, with signal processing as his area of research. His research interests include signal processing, image processing, and pattern recognition. He has authored a publication in a high impact-factor peer-reviewed journal and a paper at an international conference.

Anahita Karthik

Anahita Karthik is completing her BTech at the Symbiosis Institute of Technology in the Computer Science and Engineering department.

Basavaraj G Katageri

Basavaraj G Katageri is a Principal at the KLE Dr MS Sheshgiri College of Engineering and Technology, Belagavi. His academic qualifications include a PhD from Visvesvaraya Technological University, Belagavi, and an ME from Indian Institute of Science, Bangalore. His research areas include soil stabilization, ground improvement techniques, and environmental geo-techniques. He has several international journal and conference publications. He is a trainer for engineers of the Public Works Department, Land Army Corporation, and Power Corporation. He is a member of the Indian Geotechnical Society, Indian Society of Earthquake Engineering, Institution of Engineers (India), and Indian Society for Technical Education. He is a member of various technical advisor and academic boards of universities.

P S Kavimandan

P S Kavimandan works as an Assistant Professor at the College of Engineering, Bharati Vidyapeeth Deemed to be University, Pune. She received her BTech from Pune University and her MS from Sheffield Hallam University, UK. She is pursuing a PhD from Delhi Technological University, Delhi. She has several published papers in SCIE and Scopus indexed journals and conferences.

Paramkusham Venkata Keshava Krishna

Paramkusham Venkata Keshava Krishna received his BTech degree in electronics and communication engineering from Vallurupalli Nageswara Rao Vignana Jyothi Institute of Engineering and Technology and his MTech degree in electronics and communication engineering from Delhi Technological University, India, in

2020 and 2022, respectively. He joined the Research Laboratory as a researcher at Delhi Technological University in 2020. His research interests include signal processing, digital design, image processing, and communication systems. He has authored/co-authored three publications in international conferences and journals.

Avinash G Keskar

Avinash G Keskar was born in Nagpur, India, in 1959. He earned his BE (Hons) degree from VNIT, Nagpur, India, in 1979, his ME (Hons) degree from IISc Bangalore, India, in 1983, and his PhD from Nagpur University, India, in 1997. He is currently working as a Professor and Head of Department in the Department of Electronics and Communication Engineering, VNIT, Nagpur, India. His current research interests include computer vision, soft computing, embedded systems, and fuzzy logic. He is a Senior Member of IEEE, FIETE, FIE, and LMISTE.

Rajashri Khanai

Rajashri Khanai received her PhD in error correction coding and cryptography for wireless networks from the Visvesvaraya Technological University, Belagavi, India. Her research interests include error control codes, cryptography, and machine learning applications to signal analysis. She is currently Professor in the Department of Computer Science and Engineering, KLE's Dr MS Sheshgiri College of Engineering and Technology, Belagavi, Karnataka, India. She has published over 45 academic papers. Dr Rajashri is a senior member of IEEE.

Smith K Khare

Smith K Khare received his PhD degree in electronics and communication engineering from Indian Institute of Information Technology, Design, and Manufacturing (IIITDM) Jabalpur, India, in 2022. He is with the department of Electrical and Computer Engineering, Aarhus University, Denmark, as a postdoctoral researcher. His research interests include machine learning algorithms, artificial intelligence deployment, computer vision, real-time activity detection, real-time data and signal processing, signal denoising, time-series analysis, high performance computing, and real-time system development. He has authored/co-authored 40+ publications in various high impact-factor peer-reviewed journals. The citation impact of his publications is around 1000+ citations, with an h-index of 17 and an i10-index of 22 (Google Scholar, June 2023). He serves as a reviewer for *IEEE* transactions and reputed Elsevier journals. He has also served as a technical program committee member for several international conferences.

Amit Kukker

Amit Kukker is working with the ECE Department, Symbiosis Institute of Technology (Symbiosis International University), Pune. He has submitted his PhD thesis on 'Adaptive intelligent classification and control for biomedical and non-linear control systems' at NSIT, University of Delhi, in 2019. His research interests include biomedical signal processing, non-linear control systems and optimization techniques.

Bhagyashree V Lad

Bhagyashree V Lad was born in 1990. She received a BTech degree in electronics and telecommunications engineering from Shri Guru Gobind Singhji Institute of Engineering and Technology, Nanded, India, in 2012, and an MTech degree from Visvesvaraya National Institute of Technology, Nagpur, India, in 2015. She is currently pursuing a PhD degree with the Visvesvaraya National Institute of Technology, Nagpur, India, under the supervision of Dr A G Keskar. Her research interests include computer vision, machine learning, deep learning, and object detection.

Ruchika Malhotra

Ruchika Malhotra is Head and Professor at the Department of Software Engineering, Delhi Technological University (formerly Delhi College of Engineering), Delhi, India. She was awarded with the prestigious Raman Fellowship for pursuing postdoctoral research at the Indiana University Purdue University Indianapolis, USA. She was an Assistant Professor at the University School of Information Technology, Guru Gobind Singh Indraprastha University, Delhi, India. She received her master's and doctorate degrees in software engineering from the University School of Information Technology, Guru Gobind Singh Indraprastha University, Delhi, India. She received the IBM Faculty Award 2013. Her h-index is 22 as reported by Google Scholar. She is the author of the book *Empirical Research in Software Engineering* published by CRC Press and co-author of the book *Object Oriented Software Engineering* published by PHI Learning. Her research interests are in software testing, improving software quality, statistical and adaptive prediction models, software metrics, and the definition and validation of software metrics. She has published more than 140 research papers in international journals and conferences.

Seema Mehla

Seema Mehla received her BTech in information technology and MTech in computer science and engineering from Kurukshetra University, Kurukshetra, in 2007 and 2010, respectively. Currently, she is working as an Assistant Professor in the Department of Computer Engineering and Applications at GLA University, Mathura, India. Ms Mehla published and presented various research papers in national and international conferences. Her area of interest includes signal processing, machine learning, the Internet of Things, network security, etc.

Virender Kumar Mehla

Virender Kumar Mehla received his BTech in electronics and instrumentation engineering from Kurukshetra University, Kurukshetra, an ME in instrumentation and control engineering from National Institute of Technical Teachers Training and Research, Chandigarh, and a PhD in electrical and computer engineering from Bennett University Greater Noida, in 2004, 2011, and 2022, respectively. Currently, he is working as an Associate Professor in the Electrical and Computer Engineering Department at SRMS College of Engineering and Technology, Bareilly, India. Dr Mehla has authored various publications in good impact, peer-reviewed journals. He has written two book chapters and presented four papers at national and international conferences. His areas of interest includes signal processing, biomedical signal processing, machine learning, and the Internet of Things.

Om Mishra

Om Mishra is working at the Symbiosis Institute of Technology (Symbiosis International University) Pune. He completed his PhD at the Delhi Technological University, Delhi, an ME at the Delhi College of Engineering (Presently DTU), a Postgraduate Diploma in embedded system design at CDAC, and a BTech at UP Technical University. His research interests include signal processing, pattern recognition, and machine learning. He has published several papers in SCIE, ESCI, and Scopus indexed journals and conferences. He is a reviewer for reputed SCI-indexed international journals (IEEE transactions and Springer journals).

Rajan Mishra

Rajan Mishra is currently working as an Assistant Professor at Madan Mohan Malviya University of Technology, India. He received his PhD from Motilal Nehru National Institute of Technology, Prayagraj, India. He has published several research articles in SCI and Scopus indexed journals and conferences.

Sukumar Nagineni

Sukumar Nagineni received a BTech degree in electronics and instrumentation engineering from KITS Warangal, Telangana, India, in 2015, and an MTech degree from PDPM Indian Institute of Information Technology, Design and Manufacturing, Jabalpur, in 2018. He is currently pursuing a PhD degree with the Department of Electrical Engineering, IIT Roorkee, Uttarakhand, India. His current research interests include biomedical signal processing, digital signal and image processing, and driver monitoring and assistance systems applications.

Chandan Nayak

Chandan Nayak received a BTech degree in applied electronics and instrumentation engineering and an MTech degree in electronics and communication engineering from Biju Patnaik University of Technology, Odisha, Rourkela, India, in 2007 and 2012, respectively. He received a PhD in biomedical signal processing from the National Institute of Technology, Raipur, India, in 2021.

He was an Assistant Professor with the Department of Electronics and Communication Engineering, Bennett University, Greater Noida, India. He is working currently as an Assistant Professor in the School of Electronics Engineering, VIT-AP University, Amaravati, Andhra Pradesh, India. He has more than nine papers published in international journals and conferences. Dr Nayak has served as a reviewer for many international journals. His research interests include signal processing, biomedical signal processing, the Internet of Medical Things, fractional systems, optimization algorithms, and artificial intelligence in healthcare.

Krishna Pai

 **Krishna Pai** received a bachelor's degree in electrical and electronics engineering from KLE Dr MS Sheshgiri College of Engineering and Technology, Belagavi, India, in 2021. Currently, he serves as a Teaching Assistant in the Department of Electronics and Communication Engineering, KLE Technological University's Dr MS Sheshgiri College of Engineering and Technology, Belagavi, India. His research interests include artificial intelligence, machine learning, the Internet of Things, electrical vehicles, battery management systems, embedded systems, and communication theory. He is involved in research for numerous funded projects and the publication of related articles.

Deepak Parashar

 **Deepak Parashar** received a BE degree from Indira Gandhi Engineering College, Sagar, India, and an MTech degree in microelectronics and VLSI design from the SGSITS Indore, India. He completed a PhD degree in machine learning at Maulana Azad National Institute of Technology, Bhopal, India. He is currently working at Symbiosis Institute of Technology (Symbiosis International University) Pune. His research interests include biomedical signal processing, machine learning, and computer-aided medical diagnosis. He is an active reviewer for *The Visual Computer Journal*.

Amit Patil

 **Amit Patil** has a BTech in information technology from Jaypee Institute Of Information Technology, Noida. He has made significant contributions to the tech industry through his work as a Flutter Developer Intern at SalesBook Pvt Ltd. In less than three months, his dedication and skills helped the app grow from 10K downloads to 50K and 7000 daily active users. Patil's passion for software development extends beyond his work experience. He has also participated in Google's Adopt-a-widget program, where he began his open-source journey. He has contributed to Flutter SDK and built a package 'WhatsApp share', which gained 86% popularity on the pub.dev platform. He runs a YouTube channel where he has uploaded over 35 videos related to app development.

J Persiya

J Persiya received a BE degree from Anna University, Chennai, India, in 2009 and an ME degree from Anna University of Technology, Tirunelveli, India, in 2011. She is currently pursuing her PhD at Vellore Institute of Technology from August 2022. Her research interests include image processing.

P Prakash

P Prakash received a BE degree in electronics and communication engineering and an ME degree in communication systems from Madurai Kamaraj University, Madurai, Tamil Nadu, India, in 1998 and 2002, respectively. He was awarded with a PhD degree in information and communication from Anna University, Chennai, in 2013. He is currently working as Professor in the Department of Electronics Engineering, MIT Campus, Anna University, Chennai, Tamil Nadu, India. He has published many papers in peer-reviewed journals and conferences. His research interests include optical signal processing, digital signal processing, image processing, and communication systems.

Kemal Polat

Kemal Polat Professor Dr. Kemal Polat graduated from the Electrical-Electronics Engineering Department at Selcuk University with a B.Sc. degree in 2002 and from the Electrical-Electronics Engineering Department at Selcuk University with an M.Sc. degree in 2004. He completed his Ph.D. in Electrical and Electronic Engineering at Selcuk University in 2008. He completed his post-doctoral degree in the Department of Electrical and Computer Engineering at the University of Houston between 2015 and 2016. In his post-doctoral work, he designed mathematical modeling of memory performance by designing various experiments on visual memory. He is now working as a Professor in the Electrical and Electronic Engineering Department, Engineering of Faculty, Bolu Abant Izzet Baysal University since September 2011.

He has 185 articles published in SCI journals and 80 international conference papers. His research interests include biomedical signal classification, control systems, electronics, statistical signal processing, visual memory, neuroscience, brain-computer interface, PPG signal, medical electronics, digital signal processing, pattern recognition, and classification. His Google h-index is 52. Dr. Polat is the Associate Editor of the Information Sciences and Chaos, Solitons, and Fractals at Elsevier since 2022.

Rajkumar V Raikar

Rajkumar V Raikar is a Professor in the Department of Civil Engineering and Dean R&D, KLE Dr MS Sheshgiri College of Engineering and Technology, Belagavi, India. His academic qualifications include an MTech and PhD from Indian Institute of Technology, Kharagpur, and a BE in civil engineering. His research interests include hydraulic engineering, open channel hydraulics, fluvial hydraulics, artificial neural networks and genetic algorithms, and flow visualization using PIV and BIV. He is a member of the Editorial Board of the *Journal of Flow Measurement and Instrumentation* (Elsevier) and Academic Editor of the *Journal of Scientific Research and Reports*. He is a reviewer for 20 international journals. He has 61 international journal and 58 conference publications and has authored six books. He is Visiting Professor at National Chung-Hsing University, Taichung, Taiwan. He is a member of the International Association of Hydraulic Engineering and Research, International Association for Hydrological Sciences, Institution of Engineers India, Indian Society for Hydraulics, Indian Water Resources Society, and Indian Society for Technical Education. He is a member of the board of studies and examination of various universities and institutions.

Aditi Rao

Aditi Rao is studying for a BTech at the Symbiosis Institute of Technology in the Computer Science and Engineering department.

S Mohamed Mansoor Roomi

S Mohamed Mansoor Roomi received a BE degree, an ME degree in power systems, and an ME degree in communication systems from the Thiagarajar College of Engineering, Madurai, in 1990, 1992, and 1997, respectively, and a PhD degree in image processing from Madurai Kamaraj University in 2009. He has published more than 200 papers in national and international journals and conferences. His research interests include image processing and computer vision.

Neetu Sardana

Neetu Sardana is a Professor in the Department of Computer Science and Engineering and Information Technology (CSE/IT), Jaypee Institute of Information Technology (JIIT), Noida, India. She joined the Institute in 2012 and since then she has associated with academics and research activities at the university level. She completed her PhD at the Department of Computer Applications, Kurukshetra University, Kurukshetra, in 2011. She has more than 20 years of teaching experience. Her research interests include web mining, social network analysis, data science, and mining software repositories. She has published more than 60 research papers in various reputed indexed journals, as book chapters, and for peer-reviewed conferences. She had supervised four PhDs, fifteen MTech theses, and several BTech major and minor projects. Currently, she is guiding two PhDs. She is a member of ACM and IEEE. She is a reviewer for various journals which are of high standard and indexed in SCI/(E).

A Sasithradevi

A Sasithradevi is currently working as an Associate Professor at the Vellore Institute of Technology, Chennai, Tamil Nadu, India. She completed her ME in communication systems at Anna University and completed her PhD in the field of video retrieval from Anna University. She has published many papers in reputed journals and conferences. Her research interests include image and video analysis, pattern recognition, and machine and deep learning.

Joydeep Sengupta

Joydeep Sengupta received a BTech degree from the University of Kalyani, Kalyani, India, in 2002, an MTech degree from Burdwan University, Bardhaman, India, in 2004, and a PhD degree from the Indian Institute of Engineering Science and Technology, Shibpur, Howrah, India (formerly Bengal Engineering and Science University, Shibpur), in 2017. He is currently an Assistant Professor with the Department of Electronics and Communication Engineering, Visvesvaraya National Institute of Technology, Nagpur, India. He has authored or co-authored several research papers in reputed international journals and conferences. His research interests include microwave engineering, antenna design, optical communication, and machine learning.

Ananya Srivastava

 **Ananya Srivastava** is studying for a BTech degree at Symbiosis Institute of Technology in the Computer Science and Engineering department.

Komal Tahiliani

 **Komal Tahiliani** completed her doctoral degree in computer science engineering from Mewar University, Rajasthan. She was born and raised in Nagpur where she graduated in 2004. She has been associated with teaching computer science since 2004. She is currently working as an Associate Professor in the Computer Science and Engineering Department at Sagar Institute of Science and Technology, Bhopal. She has over 17 years of academic and research experience. Under her guidance more than 50 students have completed their postgraduate studies. Students praise her teaching skills in subjects such as C++, Python, machine learning, and data structures. She received a Best Teacher Award in 2014 for her teaching abilities. She has been involved in various activities of the Institute, such as the Internal Quality Assessment Cell, Research Committee, Special Task Force, and Institutional Development Task Force. She has organized various workshops, conferences, and faculty development programs. She has published 20+ research papers in Scopus, SCI, and other reputed national and international journals and 10+ papers in conferences. She has two patents to her name. She has delivered a number of invited talks in many institutions on topics such as cyber crime and threats, and machine learning and its application areas.

Dattaprasad Torse

 **Dattaprasad Torse** received his PhD from the Visvesvaraya Technological University, Belagavi, India, in the field of biomedical (EEG) signal analysis for seizure detection/classification applications. He received his ME from Amravati University in digital electronics. He has published over 35 research papers on EEG signal analysis in journals and conferences. He is currently Professor in the Department of Electronics and Communication Engineering, KLE Technological University, Dr MSSCET, Belagavi, Karnataka, India. He is a senior member of IEEE.

Abhuday Tripathi

 **Abhuday Tripathi** is currently working as an Assistant Professor in the Computer Science and Engineering Department at SISTEC, Gandhinagar. He received an MTech from Amity University, Uttar Pradesh, and a BTech from Maulana Abul Kalam Azad University of Technology, West Bengal (formerly West Bengal University of Technology). He has industrial experience of three years and teaching experience of more than nine years in reputed colleges. He has multiple publications in reputed journals and conferences. His academic interests are in programming, software engineering, and machine learning. He has conducted many Java and Python training sessions for students and faculties. He has also done numerous STC, FDP, and NPTEL certifications. His hobbies include travelling and writing poetry.

Deepika Varshney

 **Deepika Varshney** received her BE, MTech, and PhD degrees in computer science and engineering from Maharana Pratap College of Technology, Gwalior, Indira Gandhi Delhi Technical University for Women, and Delhi Technological University in 2014, 2017, and 2022, respectively. During her research career at Delhi Technological University (New Delhi) since 2019, she has worked on a cutting edge issue on social web platforms, fraudulent content detection on social media, and received Commendable Research Awards for publishing articles in SCI-indexed journals in 2020, 2022, and 2023. She was also awarded a Senior Research Fellowship. She is currently working as an Assistant Professor and serving academics and research in the Department of Computer Science and Engineering and Information Technology, Jaypee Institute of Information Technology, Noida, since 2022. She has also participated actively in organizing short-term training programs and other academic activities.

M Vijayalakshmi

 **M Vijayalakshmi** received a BE degree from Anna University, Chennai, India, in 2009 and an ME degree from Anna University in 2012. She is currently pursuing her PhD degree at Vellore Institute of Technology, Chennai, since January 2022. Her research interests include image processing.

Mayank Yadav

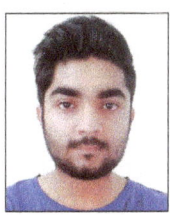**Mayank Yadav** received his BTech in computer science and engineering from Dr APJ Abdul Kalam University, Lucknow, Uttar Pradesh, India, and is pursuing an MTech degree in software engineering from Delhi Technological University, Delhi, India. He joined Delhi Technological University as a postgraduate researcher in 2021. Since August 2021 he has been with the Faculty of Software Engineering, Delhi Technological University, Delhi, where he is currently a final-year postgraduate student. His research interests include ensemble learning, virtual machines, computer vision, and cyber security. Mr Yadav has authored/co-authored three publications in peer-reviewed, Scopus, and ESCI indexed journals. Mr Yadav has presented one paper at an international conference.

IOP Publishing

Data Analytics for Intelligent Systems
Techniques and solutions
Sachin Taran, Chhavi Dhiman and Manjeet Kumar

Chapter 1

A precise ECG QRS complex detector using a WOA optimized fractional-order digital differentiator

Chandan Nayak and Manjeet Kumar

In this chapter, the design of a very accurate fractional order digital differentiator (FODD) for exact ECG QRS detection is introduced. In order to address the proposed FODD optimization problem, the whale optimization algorithm (WOA), a global optimizer, is used to estimate the near ideal solution. The WOA is equipped with an adaptive search vector to intensify and diversify the search phases separately for almost equal iterations, which not only helps it to avoid local optima but also ensures faster convergence. The pre-processor block, which is followed by peak detection logic, is implemented in the suggested QRS detection methodology using the designed optimal FODD. The suggested FODD based QRS detector performs better compared to the recently reported methods with a detection error rate of 0.86%, a positive predictivity of 99.80%, an accuracy of 99.15%, and an overall sensitivity of 99.34% when validated on PhysioNet atrial fibrillation termination challenge datasets (AFTDBs).

1.1 Introduction

Globally, cardiovascular diseases (CVDs) constitute the leading cause of death. A World Health Organization (WHO) report [1] estimates that around 17.5 million deaths worldwide each year are attributable to CVDs, which is roughly 31% of all deaths. This number will soon rise due to the stressful nature of modern living and factors such as poor physical fitness, hypertension, obesity, and dangerous alcohol and tobacco use. Importantly, CVDs, as a comorbidity, increase the risk of contracting the novel coronavirus (COVID-19). Moreover, the fatality rate of COVID-19 patients with CVD comorbidity is higher compared to non-comorbidity COVID-19 patients. In this scenario, employing the non-invasive, low-cost, and straightforward diagnostic tool called the electrocardiogram, (ECG), the fatality rate from CVDs can be decreased through early diagnosis. The P-wave, the noticeable QRS complex, and the T-wave, which are produced by depolarization of the atria, depolarization of the ventricle, and repolarization of the ventricle, respectively, make up an ECG signal.

In ECG the noticeable QRS feature makes it simpler to detect than the P- or T-waves. Identification of the QRS complex in automated ECG analysis systems is essential for diagnosing a number of potentially hazardous CVDs, including sleep apnea, myocardial infarction, heart rate turbulence, and irregular heartbeat rhythm [2, 3]. Additionally, low blood glucose levels (hypoglycaemia) in type-1 diabetic patients can be monitored continuously from the precise measurement of the R–R interval [4]. However, the task of detecting the QRS is made considerably more difficult by the irregular patterns of ECGs, which are mostly caused by artifacts and pathological states (muscular tremor, shifting of the baseline, electromagnetic interference, interference from power lines, etc). Therefore, the developed QRS detector should be resistant to artifacts and noise for correct and consistent QRS complex detection. Several researchers have tried to develop the field of QRS detection in recent years [3]. In recent times, research related to ECG signal processing continues to be carried out globally due to its frequent use to make a diagnosis of a patient's heart condition.

The pre-processor stage, which is followed by the peak identification phase, constitutes the majority of the QRS detector process [5]. Pre-processor stage tasks include suppressing noise and artifacts and creating an appropriate feature signal associated with the QRS section. The peak detection block receives the induced QRS attribute signal as an input, and either threshold-independent or thresholding-based peak detection logic is used to identify the QRS complex. In the literature, the pre-processor block is mostly implemented by employing a wavelet transform (WT) [6] and filters [7–22].

In a WT-based approaches [6] first the ECG signal is segregated into several detailed coefficients using the mother wavelet. Next, the detailed coefficients which correspond to the QRS segment are taken and further processed in the detection stage to identify the QRS complex location. Regardless of the abundant benefits of the WT-based methods, the three foremost drawbacks are as follows: (i) the absence of a mother wavelet selection rule; (ii) the selection of the mother wavelet is application-specific; and (iii) the efficacy of the technique relies upon the decomposition level of the WT.

In filter-based approaches, the filters are employed to augment the QRS segment by suppressing the noise and artifacts. In the literature, numerous filtering and QRS-related feature signal generation stages are implemented including a bandpass filter preceded by a first-order differentiator [7–14], a bandpass filter and full-wave rectifier [15], a two-stage median filter and notch filter [16], an adaptive linear predictor preceded by a Savitzky–Golay filter [17], a matched filter [18], a two-stage median filter and peak enhancement [19], a quadratic filter [20], a maximum mean minimum filter [21], and an adaptive whitening filter [22]. The enhanced QRS complexes are further actioned in the detection block for recognizing the position of the R-peak.

In the literature, peak detection logic is designed using thresholding or non-thresholding peak detection techniques. In thresholding-based peak detection logic, the enhanced QRS feature signal is compared continuously with a heuristically obtained threshold (Th) value. Whenever the envelope signal surpasses the Th limit, an R-peak has been identified, and their corresponding local maximum is considered as the R-peak location. The Th is of two types: (i) fixed Th (Th_{fixed}) and (ii) adaptive Th ($Th_{adaptive}$). The computational burden of the Th_{fixed} method is lower. However, the detection correctness of Th_{fixed} is inferior compared to $Th_{adaptive}$. Th_{fixed} logic is adopted in [20]. $Th_{adaptive}$ logic is adopted in [8–10, 17–19] and [22]. The Th based R-peak detection approaches use the duration, R–R interval, and amplitude of the previously detected QRS complex to estimate the next Th value. Thus the selection of initial parameters decides the efficiency of the peak detector.

Furthermore, in the case of missed beats, the search-back mechanism is used by considering some heuristic rules. The thresholding-based QRS detectors identify more false beats [11, 12], however, a threshold-independent QRS detector implemented using the Hilbert transform (HT) [11] is highly accurate.

Also, in recent literature, several other types of QRS detectors have been reported. A phase portrait and box-scoring-based QRS detector is reported in [23]. A parallel delta modulator-based QRS detector was reported in [24]. In [25], a nonlinear filtering scheme called relative energy is adopted to design the QRS complex detector. In [26], coordinate delay mapping is employed for QRS detection. An optimized knowledge-based fast-QRS detector is studied in [27]. A signal structural analysis based QRS detector is presented in [28].

Irrespective of the aforementioned QRS recognition methodologies, the derivative-based approach is widely employed in real-time applications, such as 24-hour Holter monitoring systems [13]. The popularity of the derivative-based approach is because of its numerous advantages, such as (i) its low cost of computation, (ii) it requires no algorithm learning period, (iii) it is independent of the patient, and (iv) it does not require manual segmentation of the ECG wave [13]. Numerous researchers have been inspired by the benefits of the derivative-based approach and have applied it to recognize the location of the ECG QRS segment. In recent literature, derivatives of both integer order [8–13] and fractional order (α) [7, 14] are applied for identifying the QRS complex location. A special case of fractional-order derivative (FOD) $D^{\alpha}x(t) = d^{\alpha}x(t)/dt^{\alpha}$ is the integer-order derivative (IOD) $D^{n}x(t) = d^{n}x(t)/dt^{n}$. Here, α and n are the real and integer number, respectively. Improved flexibility and accuracy in the designed system are attained when the differentiation of integer order is generalized to its fractional-order counterpart. Compared to the IOD, superior FOD performance is obtained for low-frequency operations [29]. Recently, the design of the FOD has grown to be a promising topic of research due to its extensive diversity of applications in several fields of engineering, including ECG QRS detection [7, 14], image segmentation [29], image texture enhancement [30], system identification [31], and transient wave propagation modeling [32]. In [7] an infinite impulse response (IIR) type fractional order digital differentiator (FODD) is realized using a conventional mathematical approach called Euler's generating function for QRS detection. In [14] a combined bandpass filter and

FODD is designed by adopting Oustaloup's forward and backward time-fractional derivative method for QRS detection. When searching the multimodal, nonuniform, and nonlinear type search terrain of the FODD design problem, the standard FODD design methodologies [7, 14] are frequently trapped in local optimal solutions. In addition, numerous FOD design methodologies have been published in the literature. Unique and advanced FOD design methodologies can be obtained from [29–32]. In general, the FOD has been designed by adopting continuous-time (CT) or discrete-time (DT) approximations [33]. Several iterative approaches such as the singularity function method [34] and least-squares method [35] have been adopted to design CT FODs. The DT FODs, called FODDs, have been designed by employing indirect or direct discretization. In the indirect discretization method, first a CT approximation is obtained for the desired frequency band, and then it is discretized [35]. The discretization process is obtained by employing a Taylor series, Maclaurin series, power series expansion, etc [36]. However, the direct discretization approach allows obtaining the FODD model directly by adopting the recursive Tustin discretization approach [37].

Discretization operator-free FODD models can also be designed by employing an evolutionary algorithm (EA). In recent time, nature-inspired EAs have found wide applications in science and technology [38–45] due to their gradient-free nature and capability of solving multidisciplinary optimization problems. In [39–42] differential evolution (DE), the real coded genetic algorithm (RGA), particle swarm optimization (PSO), and the gbest-guided gravitational search algorithm (GGSA) are used individually to optimize the FODD. The Nelder–Mead simplex algorithm (NMSA) has been employed to design a FODD [43, 44]. Due to the local search nature of NMSA, it is often stuck in a local optimum while optimizing FODD-type global optimization problems. However, EAs such as PSO and GA suffer from local optima stagnation and premature convergence which result in suboptimal solutions while tackling nonlinear, high-dimensional objective functions. To overcome these problems, in this chapter an extremely effective evolutionary whale optimization algorithm (WOA) [46] is used to design an IIR-type FODD. Due to its effective features, the WOA is widely adopted by researchers to solve numerous real-world complex optimization problems [47, 48].

Motivated by the hunting strategy of humpback whales, mathematical models are developed to make uniform global and local search phases in the WOA to ensure finding an optimal global solution with the least possible number of iterations. The local and global search phases of WOA that assist it in solving optimization problems most effectively are as follows. (i) The *global search phase*—the WOA incorporates a meticulous position updating mechanism which guarantees extensive random diversification of search agents during early stages of epochs in order to find promising regions of the search landscape. (ii) The *local search phase*—after early epochs, the WOA places high emphasis on the local search phase and faster convergence feature. During this phase, the search agents highly exploit the promising regions found during the exploration phase to arrive near the global optimum. Since in the WOA both the local and global search periods are worked out independently, for almost equal epochs the search agents cannot become trapped in a local optimum and eventually approach the global optimum faster. As a result of this, the proposed WOA-based FODD approximates the ideal half-order differentiator (HOD) magnitude response (MR) characteristic more closely over the complete frequency range. The rationale behind the present study is to demonstrate the R-peak identification applicability of the proposed FODD.

The following points summarize the key contributions of this chapter:

- An IIR-type FODD model for fractional-order, $\alpha = 0.5$ is realized with improved frequency response characteristics by employing a nature-inspired EA called the WOA.
- The designed FODDs are stable, computationally efficient, and highly accurate over the complete frequency range.
- An FODD with improved frequency response characteristics is designed in order to study its impact on QRS detection accuracy.
- The suggested QRS identifier is validated against the benchmark atrial fibrillation termination challenge database (AFTDB) ECG records.
- The suggested QRS detector provides superior metrics for positive predictivity (PP), detection error rate (DER), accuracy (Acc), and sensitivity (Se) when compared to the current techniques.

The rest of this chapter is structured as follows: the materials and methods section (section 1.2) provides a brief introduction to the metaheuristic algorithm, articulates the proposed IIR-type FODD design problem analytically, describes the WOA briefly, explains the proposed QRS detection methodology, and also provides a brief description of the employed ECG database. The results section (section 1.3) illustrates total and comprehensive simulated outputs and comparisons with published articles. Finally, section 1.4 concludes the overall outcomes of the proposed approach.

1.2 Materials and methods

1.2.1 Introduction to the metaheuristic algorithm

Metaheuristic algorithms inspired by nature solve problems of optimization by mimicking physical or biological phenomena. Broadly, they are clustered into three key classes: evolution-based, physics-based, and swarm-based. Swarm-based approaches emulate the social interactions of animal communities. The most widely recognized swarm-based approaches are the cuckoo search algorithm (CSA), salp swarm algorithm (SSA), ant colony optimization (ACO), PSO, WOA, etc. Physics-based approaches mimic the Universe's physical laws. The gravitational search algorithm (GSA), ray optimization (RO), and black hole algorithm (BHA) fall into this category. Lastly, the rules of natural evolution inspire evolution-based approaches such as the biogeography-based optimizer (BBO), genetic programming (GP), evolution strategy (ES), and GA. Despite differences between the swarm-based, physics-based, and evolution-based techniques, the universal aim is to improve a single or a set of solutions during optimization. Algorithm 1 describes a comprehensive schema that forms the solid framework for all the numerous variants of metaheuristic algorithms in order to present a unified view. The general engineering behind all these approaches is identical: provided a population of individuals with limited resources within a certain environment, the contest for these resources induces natural selection (survival of the fittest). This increases or decreases the population's fitness for maximization or minimization problems, respectively. For a fitness function to be maximized (or minimized), initially random initialization of candidate solutions takes place. The fitness of each candidate solution is found. A higher fitness of the search agent is considered better for a maximization problem, whereas a lower fitness of the search agent is assumed to be better for a minimization problem. Thus, the fitness value is used to elect the best search agent (B_{search_agent}) from the entire population. After that a metaheuristic algorithm-driven random search (the name of the random search depends upon the algorithm employed) is carried out in each epoch to amend the position of the entire population. The fitness of each amended candidate solution at a new position in the search area is evaluated. Then, the fitness of B_{search_agent} is compared with the fitness of each candidate solution and the B_{search_agent} is replaced by the candidate solution which has better fitness compared to the cost of B_{search_agent}. This procedure is continued until a B_{search_agent} with adequate quality is obtained or the computational limit previously set is attained. Finally, the position of the B_{search_agent} is used as the optimal result for the given optimization problem.

Algorithm 1:. General pseudo-code of the metaheuristic algorithm

---

Begin
1: Initialize populations with random solutions for candidates
2: Estimate each and every candidate
3: Find b_{search_agent}
4: **Reiterate** until (end criteria are assured) **do**
5:      Amend the position of each candidate according to the algorithm-driven random search
6:      Estimate amended candidates
7:      Update b_{search_agent} if *fitness* (b_{search_agent}) < *fitness* (amended candidate)
8: **End Reiterate**
End Begin

1.2.2 Proposed fifth-order FODD design

The frequency response (FR) of an ideal FODD $H_{IDD}^{\alpha}(Z)$ is defined by

$$H_{IDD}^{\alpha}(\omega) = |\omega|^{\alpha} < 90° \times \alpha, \qquad (1.1)$$

where α is the fractional order of the FODD and $\omega \in [0, 1]$ is the normalized frequency.

The transfer function (TF) of an Nth-order IIR filter is defined by

$$H_{\text{DD}}(Z) = \frac{A(Z)}{B(Z)} = \frac{\sum\limits_{i=0}^{N} a_i Z^{-i}}{\sum\limits_{i=0}^{N} b_i Z^{-i}}, \tag{1.2}$$

where b_i and a_i, for $i = 0, 1, 2, 3, \ldots, N$, are coefficients of $H_{DD}(Z)$ and N is the order of the FODD.

Here, $N = 5$ and $\alpha = 0.5$ are considered to achieve low computational complexity and superior QRS recognition correctness, respectively, as examined in the following subsections.

The FR of $H_{DD}(Z)$ is obtained by putting $Z = e^{j\omega}$. It is defined as

$$H_{\text{DD}}(\omega) = \frac{\sum\limits_{i=0}^{N} a_i e^{-j\omega t}}{\sum\limits_{i=0}^{N} b_i e^{-j\omega t}}. \tag{1.3}$$

Figure 1.1 shows a schematic of the IIR-type FODD design system. The goal is to design an optimized coefficient vector $C_v = [a_i, b_i]$ such that its FR characteristic closely approximates to the ideal FODD counterpart $H_{\text{IDD}}^{\alpha}(Z)$. To achieve this, in each epoch of the WOA the maximum absolute magnitude error (MAME) $J(a_i, b_i)$ between $H_{\text{IDD}}^{\alpha}(\omega)$ and $H_{\text{DD}}^{\alpha}(\omega)$ is calculated for $L = 1024$ sample instants:

$$J(a_i, b_t) = \max(||H_{\text{IDD}}^{\alpha}(\omega)| - |H_{\text{DD}}(\omega)||). \tag{1.4}$$

Next, the fitness MAME value is feedback to the WOA to alter the coefficient vector C_v in such a way that $J(a_i, b_i)$ is minimized to its lowest error value. The coefficient

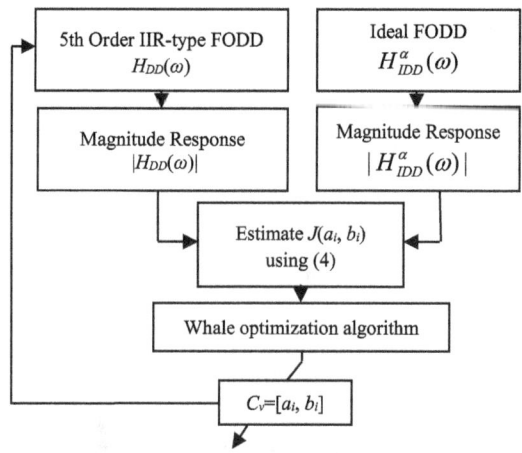

Figure 1.1. The proposed WOA-based FODD design methodology.

vector, C_v corresponding to the minuscule error cost of $J(a_i, b_i)$ is used as the best possible result of the proposed FODD design problem.

The average group delay (τ_{DD}) metric is defined by

$$\tau_{DD}(\omega) = \frac{1}{\pi} \int_0^{\pi} \tau_D(\omega) d(\omega), \tag{1.5}$$

where $\tau_D(\omega) = -d\theta_{DD}(\omega)/d\omega$ $\theta_{DD}(\omega) = \angle H_{DD}(\omega)$ is the phase response of the $H_{DD}(\omega)$. It is used to assess the phase characteristic of the proposed FODD. To avoid undesired phase distortion, the phase characteristics of the realized FODD must be linear.

1.2.3 Whale optimization algorithm (WOA)

The WOA is a nature-inspired EA that is based on the ingenious bubble-net hunting stratagem of the whales [46]. The WOA applies an evenly distributed control mechanism to enable simultaneous local exploitation and global exploration of the search landscape. This innovative control mechanism of the WOA amplifies the search capability and circumvents premature convergence by escaping from local optima to arrive closer to the global best solution. Whales like to pursue and kill shoals of small fish (prey) adjacent to the surface of the sea. To hunt a shoals of small fish, the whales initially search aimlessly for shoals of small fish in the sea. Once the whereabouts of the prey is identified, the whales surround the prey and hunt it by swimming up towards the surface of the sea around the prey within a shrinking circle by blowing binding bubbles along a circular path. To perform the optimization, three effective mechanisms, namely encircling the prey, bubble-net feeding, i.e. the exploitation mechanism, and the search for prey, i.e. the exploration mechanism, are mathematically modeled in the WOA, as discussed in the subsequent subsections.

1.2.3.1 Encircling the prey
In the WOA the locations of the whales (search agents) are randomly initialized in the search landscape as defined by

$$X_i = \left(x_i^1, x_i^2, x_i^3, \cdots, x_i^d, \cdots x_i^D \right) \text{ for } i = 1, 2, 3, ..., n_p, \tag{1.6}$$

where n_p, D, and x_i^d are the total number of whales, the total number of search landscape dimensions, and the location of the ith humpback whale in the dth dimension of the search landscape, respectively.

The pursuing strategy of the whales is initiated by encircling the prey. The encircling mechanism is mathematically modeled as

$$D = |X_*(t). \, C - X(t)| \tag{1.7}$$

$$X(t+1) = X^*(t) - D. \, A, \tag{1.8}$$

where X^* and X are the best and the most recent position vectors of the humpback whales, respectively. The coefficient vectors A and C are defined by, respectively,

$$A = 2. \, r. \, a - a \tag{1.9}$$

$$C = 2. \, r, \tag{1.10}$$

where $r \in (0, 1)$ denotes an arbitrary number, and the value of a is diminished from 2 to 0 throughout the epoch. The humpback whales can be positioned around the current best solution (target prey) by varying the values of C and A.

1.2.3.2 Bubble net feeding: the exploitation mechanism

In the WOA, the spiral updating and shrinking encircling mechanisms are independently implemented to realize the mathematical design of the whales' bubble-net feeding mechanism. The shrinking encircling strategy is implemented by diminishing the cost of a in equation (1.9). The helix shape maneuver of the humpback whales is mathematically modeled to implement the spiral updating mechanism:

$$X(t + 1) = X_*(t) + D'. \, e^{bl}. \, \cos(2l\pi), \tag{1.11}$$

where $l \in [-1,1]$ represents the uniform arbitrary number and $D' = |X_*(t) - X(t)|$ is the space between the ith the whale and prey. The logarithm spiral shape constant is represented by b.

During hunting, the humpback whales move in both the spiral-shaped and shrinking encircling path simultaneously. To simultaneously execute both maneuvers of the whale, a random parameter $p \in [0,1]$ is employed to select either of the maneuvers of the humpback whale by using

$$X\left(t + 1\right) = \begin{cases} (X_*(t) - A. \, D) & \text{if } (p < 0.5) \\ D'. \, e^{bl}. \, \cos(2\pi l) + X_*(t) & \text{if } (p \geqslant 0.5) \end{cases}. \tag{1.12}$$

1.2.3.3 Search for prey: the exploration mechanism

In the exploration stage, the humpback whales amend their locations by using a randomly selected humpback whale rather than using the best humpback whale. Also, to intensify the search process, the value of $|A| > 1$ is used. The mathematical models of the exploration mechanism are given by

$$D = |X_{\text{rand}}C - X(t)| \tag{1.13}$$

$$X(t + 1) = X_{\text{rand}} - D. \, A, \tag{1.14}$$

where X_{rand} is the randomly selected humpback whale.

The WOA steps for designing the IIR-type fifth-order FODD are as follows.

Step 1. For the Nth-order FODD design, the positions of n_p ($= 100$) humpback whales are randomly initialized in the D ($= 2 \times (N + 1)$) dimensional problem landscape using equation (1.6). Here, each whale comprises $N + 1$ denominator (b_i) and numerator (a_i) coefficients, for $i = 1, 2, ..., N + 1$. Utilizing equations (1.9) and

(1.10), calculate the coefficient vectors A and C, respectively. Initialize the value of the logarithm spiral shape constant $b = 1$.

Step 2. By using equation (1.4), calculate the fitness value $J(a_i, b_i)$ of each humpback whale. Declare the leader whale X^* which has the lowest $J(a_i, b_i)$ value.

Step 3. For each humpback whale, update the control variables A and C by using equation (1.9) and (1.10), respectively. Initialize the values of the other control variables l and p with randomly generated values.

Step 4. If $p < 0.5$ then execute either step 5 or step 6, otherwise perform step 7.

Step 5. By using equation (1.8) amend the location of the humpback whale if $|A| < 1$.

Step 6. By using equation (1.14) amend the location of the humpback if $(|A| > 1)$.

Step 7. By using equation (1.11) amend the location of the humpback whale if $(p \geqslant 0.5)$.

Step 8. Reallocate the location of the humpback whale within the upper (ub = 10) and the lower (lb = −10) search bound of the design variable when any humpback whale moves out of it.

Step 9. Utilizing the formula given in equation (1.4), determine $J(a_i, b_i)$ for each humpback whale. For an improved outcome, update the leader whale X^*.

Step 10. Execute steps 4 to 10 until t_{max} is reached.

Step 11. After t_{max}, return the location of the leader whale X^* as the optimal result of the FODD.

Figure 1.2 shows a flow diagram of the WOA-based FODD design.

1.2.4 Proposed QRS complex identification scheme

Real-time use of the designed FODD is demonstrated by utilizing it in the QRS recognition application. The structural design of the suggested QRS finder is explained in figure 1.3(A) [11, 49]. It incorporates four stages: (i) noise elimination, (ii) the production of marker signal-related QRS features, (iii) augmentation of the generated feature, and (iv) R-peak identification. The working of each block is described in figure 1.3(B).

1.2.4.1 BPF-based noise elimination

In practice, numerous artifacts and noise sources, such as muscle tremors, shifting of the baseline, electromagnetic interference, interference from the power line, etc, contaminate the ECG signal. The presence of these unwanted ECG components hinders the QRS recognition task. Since the majority of the QRS complex energy is contained within the 5–22 Hz frequency region, a (5–22 Hz) tenth order Butterworth BPF is applied to the raw ECG signal (see figure 1.3(B)(a)) to discard the undesired ECG regions. The bandpass-filtered ECG signal is presented in figure 1.3(B)(b).

1.2.4.2 Proposed FODD based QRS-related feature signal generation

In the derivative-based methods, the slope information created by the differentiator is considered a marker corresponding to each R-peak. This induced slope information is further actioned in the detection stage to identify the real location of QRS. To

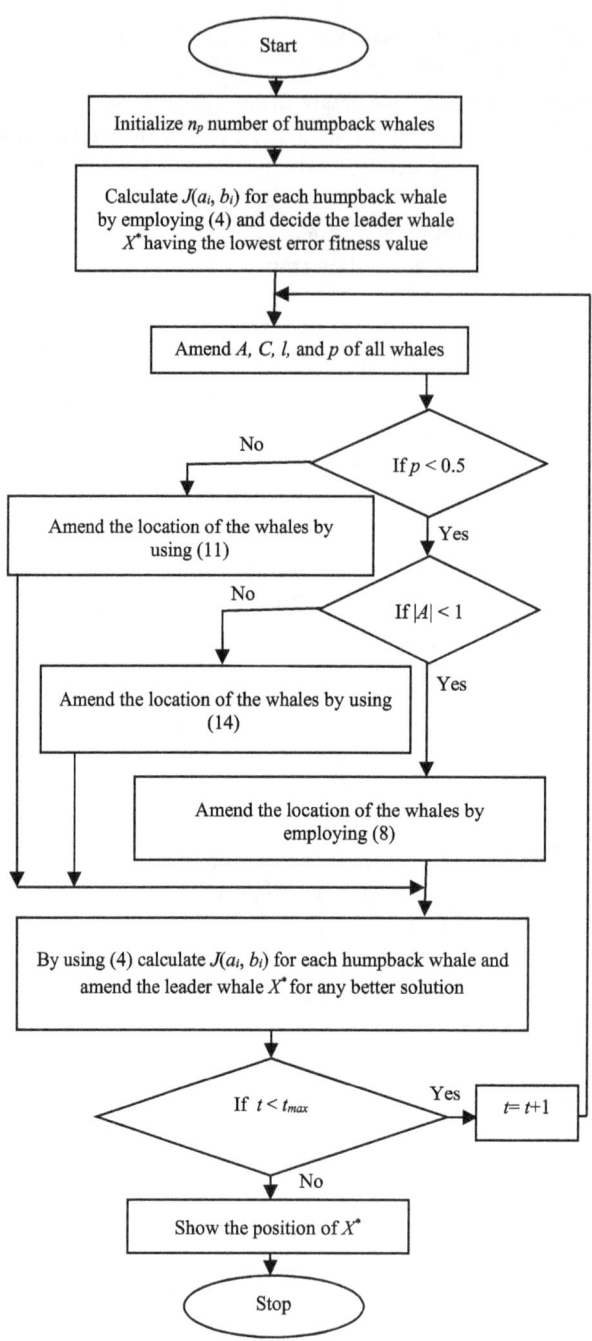

Figure 1.2. Flow diagram of the WOA-based FODD design problem.

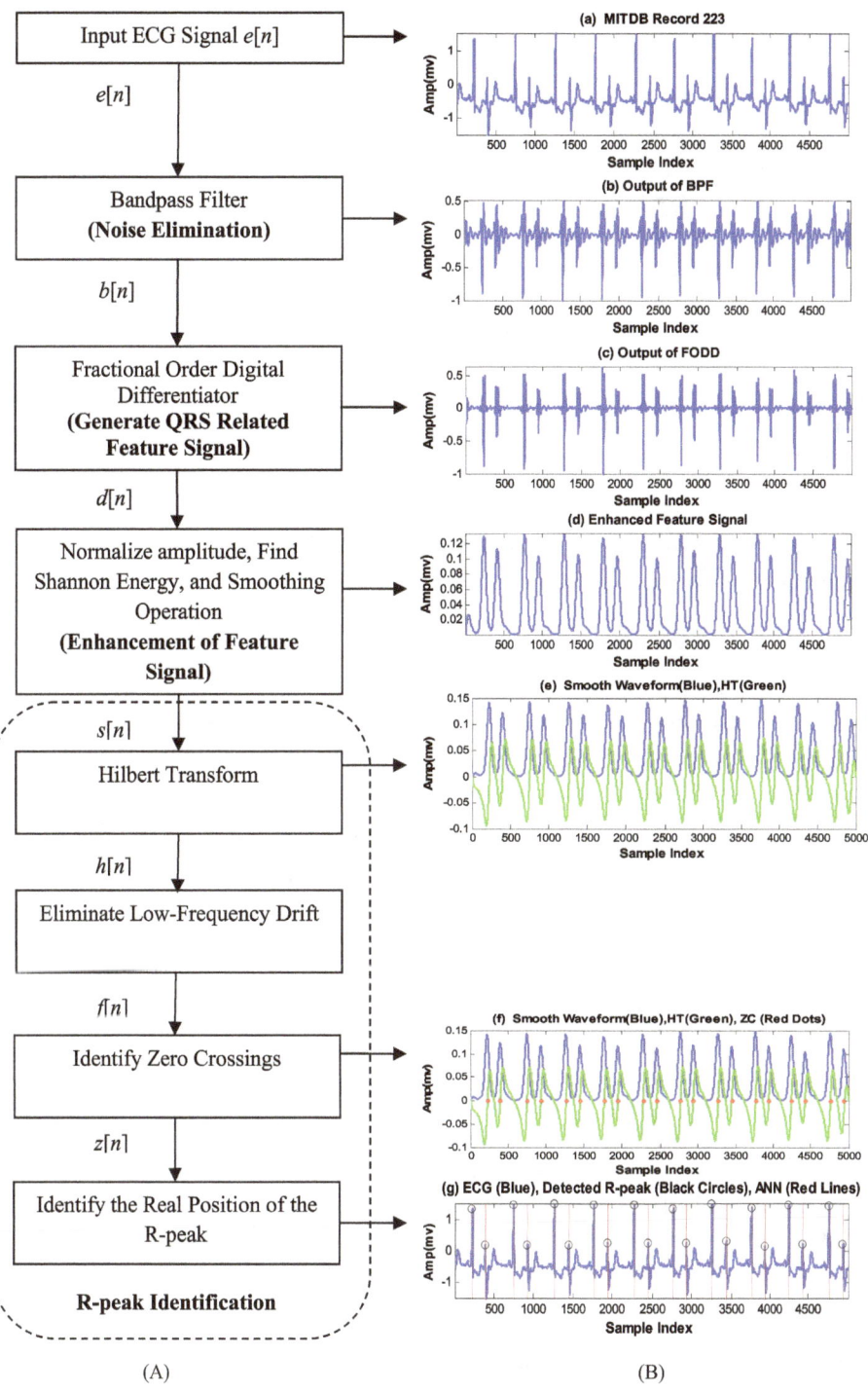

(A)   (B)

Figure 1.3. (A) QRS complex identification scheme. (B) Illustration of the proposed QRS complex identification scheme: (a) ECG record 223 of MITDB, (b) output of BPF, (c) QRS-related feature signal generated by proposed FODD, (d) enhanced (smooth) feature signal, (e) smooth envelope and o/p of HT, (f) smooth envelope, o/p of HT, and ZCs, (g) identified R-peaks (black circle) and database annotation (ANN) (red standing lines).

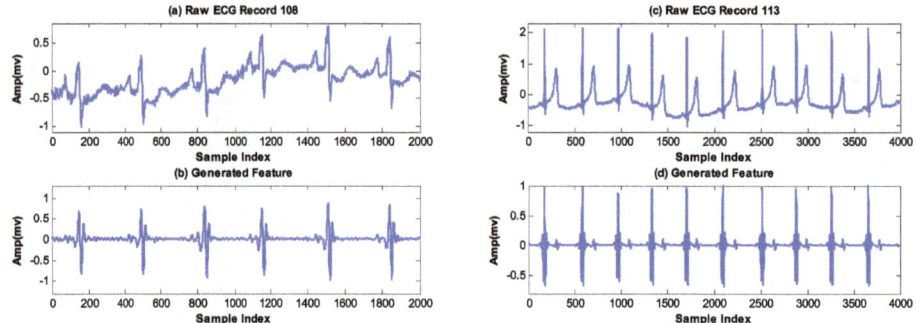

Figure 1.4. An example of the generated feature signal, $d[n]$, for various aberrant ECG fragments taken from the MITDB: (a) ECG beats showing a first-degree atrioventricular block, (b) generated feature, (c) ECG with high amplitude T-wave, and (d) generated feature.

extract the QRS segment feature, the bandpass processed ECG wave, $b[n]$ is passed through the proposed FODD, $H_{\text{DD}}(Z)$, as shown in

$$H_{\text{DD}}(z) = \frac{a_0 + a_1 z^{-1} + a_2 z^{-2} + a_3 z^{-3} + a_4 z^{-4} + a_5 z^{-5}}{b_0 + b_1 z^{-1} + b_2 z^{-2} + b_3 z^{-3} + b_4 z^{-4} + b_5 z^{-5}}, \qquad (1.15)$$

where the numerator coefficients are $a_0 = 1.0773$, $a_1 = -3.0651$, $a_2 = 2.8649$, $a_3 = -0.7175$, $a_4 = -0.2725$, and $a_5 = 0.1128$, and the denominator coefficients are $b_0 = 1.0112$, $b_1 = -2.2616$, $b_2 = 1.3723$, $b_3 = 0.1145$, $b_4 = -0.2682$, and $b_5 = 0.0343$ for the proposed FODD. The WOA is used to determine the values of these denominator and numerator coefficients, which are described in the following sections. Figure 1.3(B)(c) shows the generated feature wave. From figure 1.3(B)(c) it can be seen that the convolution process once more minimizes the influence of the T/P-waves. Additionally, numerous anomalous ECG segments are visually examined to ensure the strong feature generation competence of the suggested FODD. Figure 1.4 shows two irregular ECG fragments as an example. The ECG beats of figure 1.4(a) show a first-degree atrioventricular (AV) block. For the adult population, the PR interval lies between 120 and 200 msec. However, the first-degree AV block shows a prolonged PR interval (>200 msec) on an ECG. Figure 1.4(c) shows elevated T-waves. Clinically the elevated T-wave shows coronary artery blockage. From the generated feature wave as illustrated in figures 1.4(b) and (d), it can be seen that the suggested FODD has the ability to generate feature signals in all of the atypical circumstances stated above. The subsequent blocks further process the generated feature signal to identify the R-peak locations.

1.2.4.3 QRS feature enhancement
To make the process of QRS detection easier, the proposed feature enhancement stage is used to provide a single smooth envelope corresponding to the generated feature signal. For this, amplitude normalization of the differentiated ECG signal, d

$[n]$, is first performed by employing the following equation to further reduce the undesired ECG regions:

$$a[n] = \frac{d[n]}{\max\limits_{n=1}^{k}(|d[n]|)},\qquad(1.16)$$

where K is the ECG signal sample.

The amplitude-normalized signal, $a[n]$ is a bipolar signal. To identify the inverted QRS complex quickly, in this chapter a Shannon energy transform (SET) is employed. The SET performs nonlinear amplification to the amplitude-normalized signal, $a[n]$ using

$$e_s[n] = -(\log (a [n])^2 . (a [n])^2).\qquad(1.17)$$

It has been observed that the generated Shannon energy envelope is not smooth enough, and it includes multiple peaks. Hence, it is further smoothed by employing a smoothing filter of moving average (MA) type, as shown in

$$s[n] = \frac{1}{M} \sum_{m=0}^{M-1} e_s[n - (M - 1 - m)],\qquad(1.18)$$

where $M (= 55)$ is the MA filter's length.

Figures 1.3(B)(c) and (d) show the differentiated feature signal, $d[n]$, and the smoothed envelope, $s[n]$, respectively. It can be observed that the local maximum in each smooth wave, $s[n]$, represents the rough ECG R-peak position. For this reason, the local maximum related to every envelope wave is initially identified by utilizing a non-amplitude thresholding method called HT. Subsequently, the local maximum is engaged to identify the correct R-peak position in the ECG wave.

1.2.4.4 R-peak identification

The R-peak identification module is the cascaded connection of four signal processing blocks [49]: (i) HT, (ii) low-frequency (LF) shift remover, (iii) zero crossing (ZC) identifier, and (iv) accurate R-peak identifier. The maneuver of each block is explained below.

Let $r(t)$ be a practical signal and $i(t) = 1/\pi t$ depict the HT's impulsive response. $\hat{r}(t)$ is the HT of $r(t)$:

$$\hat{r}(t) = HT [r(t)] = \left[\frac{1}{\pi t} * r(t)\right] = \frac{1}{\pi} \int_{-\infty}^{\infty} \frac{r(\tau)}{t - \tau} d\tau.\qquad(1.19)$$

By utilizing the Fourier transform's (FT's) convolution characteristic one obtains

$$F[\hat{r}(t)] = F[i(t) \times F[r(t)]\qquad(1.20)$$

$$\hat{R}(f) = I(f)R(f)\qquad(1.21)$$

$$\hat{R}(f) = F[r(t) \times F\left[\frac{1}{\pi t}\right] = -R(f)j \, \text{sgn}(f). \tag{1.22}$$

Here $\text{sgn}(f) = \begin{cases} +1 & f > 0 \\ 0 & f = 0 \\ -1 & f < 0 \end{cases}$.

As a result, the HT of $r(t)$ can be determined using

$$\hat{r}(t) = \text{IFT}[\hat{R}(f)] \text{ where } \hat{R}(f) = \begin{cases} jR(f) & \text{if } f < 0 \\ -jR(f) & \text{if } f > 0 \end{cases}. \tag{1.23}$$

Here IFT stands for inverse FT and $\hat{R}(f)$ represents the FT $r(t)$.

The HT-based local maxima finding logic is illustrated in figure 1.3(B)(e). The HT (see the green line in figure 1.3(B)(e)) of the local maximum (see the blue line in figure 1.3(B)(e)) is calculated. It can be observed that due to the odd symmetry property of HT, the HT output crosses the zero reference line corresponding to the inflexion point in each local maximum. As a result of which the ZC positions (see the red dots in figure 1.3(B)(f) and their corresponding local maximum fall on similar perpendicular lines. The position of the ZCs is found by employing the ZC identifier block.

However, for the successful detection of the ZC locations, the HT sequence should align appropriately over the reference (ZC) line. But, due to variable ECG morphology, often it is observed that the output of HT deviates from the ZC line, known as low-frequency (LF) shift. The number of false negative (FN) beat counts rises as a result of the shifting of HT output from the ZC line. Therefore, the QRS recognition algorithm should avoid this uncommon LF shift issue. In order to stabilize the HT output over the ZC line, first the moving average (MA) filter's response to the HT input is obtained then the response of the MA filter is subtracted from the original HT output. This process is carried out by the LF shift remover block. Here, the MA filter's length plays a critical role in eliminating the LF shift issue. The length of the MA filter is established experimentally. In this work, experimental results show that $K_{ma} = 400$ is the length of the MA filter for ECG waves with a 360 Hz sampling frequency.

There is a microscopic dissimilarity between the time of the incident of the R-peaks and the ZC locations. The accurate incident of the R-peaks is obtained by adopting the real R-peak identifier, where the real incidences (refer to figure 1.3(B) (g)) of the R-waves are detected by exploring the absolute maxima in the input ECG wave contained by ±25 samples nearby each ZC locations.

1.2.5 ECG databases and the performance measurement

Globally, researchers validate arrhythmia detectors by employing the Physionet ECG datasets [50]. The suggested QRS detector is visually investigated using the MIT/BIH Arrhythmia Database (MITDB) [51] and thoroughly verified utilizing the full upper channel AFTDB ECG datasets [52]. The MITDB is made up of dual channel 48 ECG recordings from patients with various diseases, recorded at a rate of 360 Hz and with a 11-bit resolution over a ±5 mV range [51]. Each signal in the AFTDB consists of 60 s atrial fibrillation segments with two ECG signals that have each been collected at a rate of 128 samples per second [52]. Each ECG record of the AFTDB is connected with an annotation file. The exact R-peak location can be found in these annotation files.

The false negative (FN), false positive (FP), and true positive (TP) beats are chosen using the annotation file as a guide. The total number of missed, erroneously detected, and accurately detected heartbeats are shown here by the letters FN, FP, and TP, respectively. The suggested R-peak detector performance is validated using the following performance evaluation criteria:

$$Se = TP/(FN + TP) \tag{1.24}$$

$$PP = TP/(FP + TP) \tag{1.25}$$

$$DER = (FN + FP)/(\text{Total number of QRS complexes}) \tag{1.26}$$

$$Acc = TP/(TP + FP + FN). \tag{1.27}$$

1.3 Results

All the simulations were performed in the MATLAB environment, run on an Intel core i5 9th generation processor and Windows 10 operating system.

1.3.1 Simulation results of the proposed FODD

1.3.1.1 FR analysis of the proposed FODD
The proposed WOA-based FODD plays a crucial role in recognizing the incidence of accurate R-waves. The order of the IIR filter, N is considered as 5. Extensive MATLAB simulations are performed for the proposed filter design problem. Table 1.1 presents the best control variables of the WOA, which are selected by running the algorithm several times. The optimal sets of numerator coefficient, denominator coefficient, and the estimated values of the magnitude response error metrics (root mean square magnitude

Table 1.1. Control variables of WOA for the realization of the proposed FODD.

Control variable	WOA
Size of the population (n_p)	100
Maximum number of epochs (t_{max})	400
Logarithmic spiral shape constant (b)	1
Iteration-dependent control variable (a)	Varies from 2 to 0
Random parameters (r and p)	[0,1]
Random parameter (l)	[−1,1]

Table 1.2. TF of the proposed WOA-based FODD of fifth order and its performance metrics.

A	$H_{DD}(z) = \dfrac{a_0 + a_1 z^{-1} + a_2 z^{-2} + a_3 z^{-3} + a_4 z^{-4} + a_5 z^{-5}}{b_0 + b_1 z^{-1} + b_2 z^{-2} + b_3 z^{-3} + b_4 z^{-4} + b_5 z^{-5}}$	RMSME	RMSME (dB)	τ_{DD} (samples)
0.5	$H_{DD}(z) = \dfrac{1.0773 - 3.0651 z^{-1} + 2.8649 z^{-2} - 0.7175 z^{-3} - 0.2725 z^{-4} + 0.1128 z^{-5}}{1.0112 - 2.26161 z^{-1} + 1.3723 z^{-2} + 0.1145 z^{-3} - 0.2682 z^{-4} + 0.0343 z^{-5}}$	0.0068	−43.29	1.1647

error (RMSME) $= \sqrt{\dfrac{1}{L}\displaystyle\sum_{l=1}^{L}\||H_{\text{IDD}}^{\alpha}(\omega)| - |H_{DD}(\omega)|\|^2}$ and average group delay (τ_{DD})) of the proposed WOA-based designed FODD are presented in table 1.2. Figure 1.5(a) depicts the magnitude response (MR) of the proposed WOA-based FODD. In figure 1.5(a), it can be seen that the magnitude response of the proposed WOA-based FODD approaches very accurately the magnitude response of its corresponding ideal differentiator counterpart for the entire normalized frequency band. The absolute relative magnitude error (ARME $= |\{(|H_{\text{IDD}}^{\alpha}(\omega)| - |H_{HDD}(\omega)|)/(H_D^{\alpha}(\omega)|\}|$) and absolute magnitude error (AME $= \||H_{\text{IDD}}^{\alpha}(\omega)| - |H_{DD}(\omega)|\|$) between $H_{\text{IDD}}^{\alpha}(\omega)$ and $H_{DD}(\omega)$ are evaluated for $L = 1024$ sample instants and are shown in figures 1.5(b) and (c), respectively. Figure 1.5(d) shows the group delays response plots of the proposed WOA-based FODD. In figure 1.5(d), it can be observed that the group delay plot is almost constant for $\omega \geqslant 0.02\pi$, and with a low τ_{DD} value of 1.1647 samples, as reported in table 1.2. The aforementioned desirable features make the proposed WOA-based FODD appropriate for the ECG QRS identification.

1.3.1.2 Average performance study of the proposed FODD through the RMSME (dB) metric

The WOA-based FODD algorithm was run 25 independent times to examine the average case performance statistics of the proposed design in terms of the RMSME (dB) metric. The resulting statistical metrics such as best, worst, mean, median, standard deviation (SD), and variance values are −43.29 dB, −42.88 dB, −43.21 dB, −43.24 dB, 0.071, and 0.0050, respectively, as shown in table 1.3. The SD value of 0.071 in table 1.3 confirms the robustness of the WOA for solving the proposed FODD-type optimization problem.

1.3.1.3 Stability analysis

The pole-zero graph shown in figure 1.6(a) shows that the location of all poles is contained by the unit circle and thereby confirms the stability of the proposed WOA-based FODD.

1.3.1.4 Convergence analysis

If an algorithm is capable of avoiding the local optima, it does not guarantee that the solution will proceed toward the global optimum. In general, a possible rough

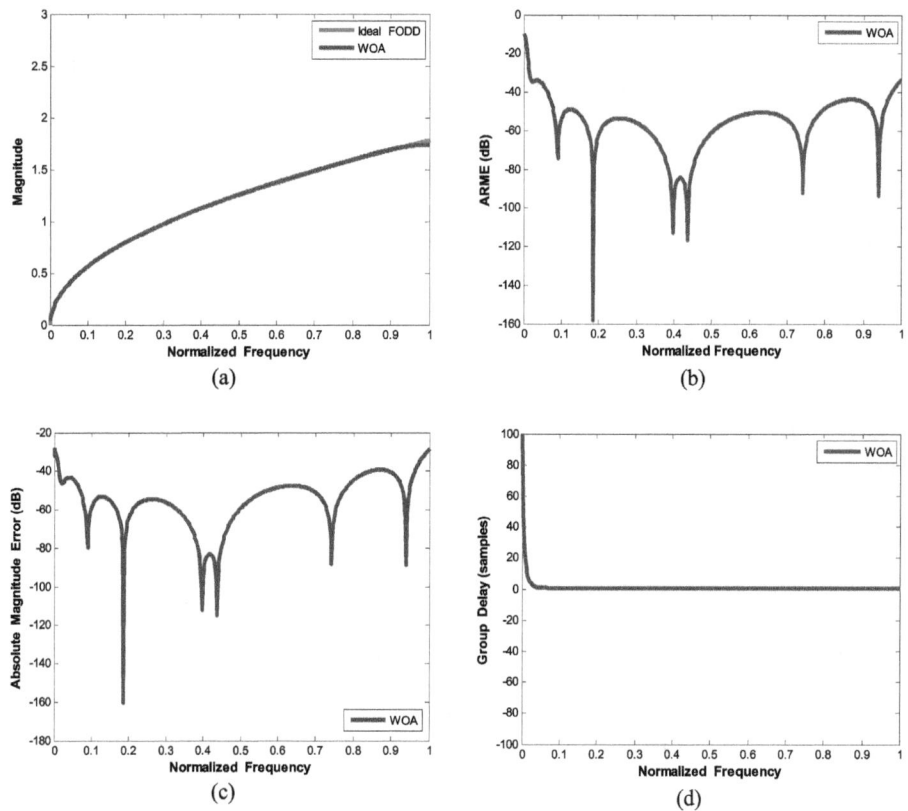

Figure 1.5. FR plots of the proposed WOA-based FODD: (a) MR, (b) ARME plot, (c) AME plot, and (d) group delay response.

Table 1.3. Statistical data of RMSME (in dB) for the proposed WOA-based FODD.

Error metric	Best	Worst	Mean	Median	SD	Variance
RMSME (dB)	−43.29	−42.88	−43.21	−43.24	0.071	0.0050

solution is found when an algorithm evades the local optima. Then the algorithm should try to enhance the accuracy of the obtained rough solution. The convergence speed tells about the rate at which an EA proceeds toward the global optimum. Figure 1.6(b) shows the convergence profile of the WOA-based FODD design problem. From the convergence plot, it can be visualized that the WOA attains the least possible error fitness value of 0.149 in 209 epochs. To establish the robustness of the adopted algorithm, 25 independent trial runs are performed, and various statistical parameters are presented in table 1.4. The box plot is also drawn in figure 1.4(c) to emphasize the importance of this experiment.

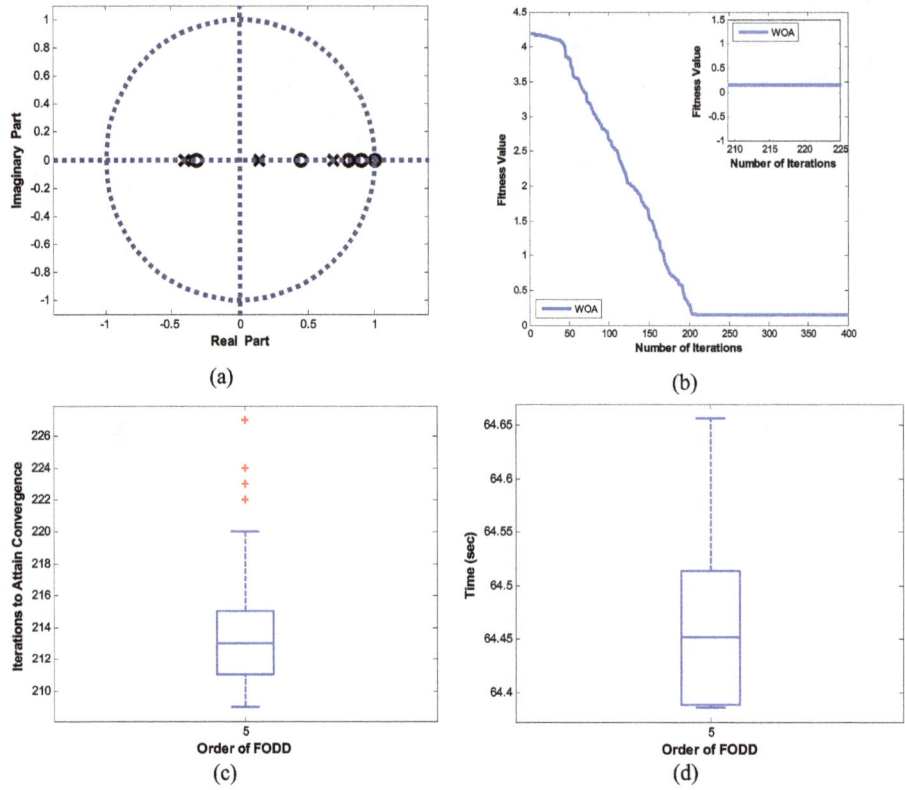

Figure 1.6. The response of the proposed WOA-based FODD: (a) the pole-zero diagram, (b) convergence profile, (c) box plot showing the convergence of the WOA for the design of FODD, and (d) run time required by the WOA for the FODD design using a box plot.

Table 1.4. Convergence profile and execution time statistical analysis for the WOA-based FODD algorithm.

Statistical metrics	Number of epochs needed to attain the lowest error cost value	Execution time (s) for 400 iteration cycles
Minimum	209	64.3860
Maximum	227	64.6560
Mean	213.54	64.4555
Median	213	64.4520
SD	4.086	0.0703
Variance	16.69	0.0049

1.3.1.5 Run time investigation

In this section, the exploring efficiency of the WOA for the realized FODD-type design problem is studied in terms of the average execution time. For this, a statistical experiment for 25 individual runs has been carried out, and the obtained outcomes are presented in table 1.4 and figure 1.6(d). The acceptable mean execution time of 64.4555 s is required for running the proposed WOA-based FODD algorithm for 400 iteration cycles.

1.3.2 Sensitivity analysis

The final solution quality resulting in any optimization algorithm depends on the accurate tuning of the algorithm's internal control variables. In the WOA, the internal control variables are the population size (n_p), logarithmic spiral shape constant (b), and iteration-dependent control variable (a). To find out the best internal control variable values, the WOA is run 25 independent times with a variation of each control variable, and the values of the control variables which result in the best performance in terms of the RMSME index are selected. During the selection of an individual control variable value, the rest of the control variable values remain unchanged, as shown in table 1.1.

1.3.2.1 Sensitivity analysis due to the variation of n_p
For the WOA-based FODD design problem, the result is dependent on the exploitation and exploration capabilities of the WOA. The size of the population directly influences the exploration efficiency of the algorithm. The impact of the population size variation on the quality of the end solution in terms of the RMSME (in dB) metric of the designed FODD is studied by considering 50, 100, and 150 humpback whales, and the corresponding solutions are shown in table 1.5. The results illustrated in table 1.5 ensure that by considering a smaller number of humpback whales, $n_p = 50$, the search area is not explored thoroughly, which results in a low-quality solution. At the same time, by considering more humpback whales, $n_p = 150$, though the search landscape is efficiently explored, however, no significant enhancement in the RMSME metric is observed as compared to 100 number of humpback whales. Also, for 150 number of search agents, the algorithm execution time is incremented unnecessarily. Thus, by considering the trade-off between the execution time and the quality of the end solution, it can be inferred that 100 number of humpback whales are the most suitable choice for the proposed FODD design.

1.3.2.2 Sensitivity study because of the variation of b
In the WOA the spiral movement of the humpback whales is controlled by the logarithmic spiral shape constant b. For a very high value of b, the spiral tends towards a straight half-line, and for a small value of b, the spiral becomes a circle. The consequences due to the variation of the initial value of the logarithmic spiral shape constant b are studied in this section, and the outcomes are shown in table 1.5. From table 1.5, it is confirmed that the better RMSME (dB) metric is achieved for $b = 1$.

1.3.2.3 Sensitivity analysis due to the variation of a
In the WOA balanced local and global search phases are maintained by employing the iteration-dependent control variable a. The effect due to the change in a is studied in this section, and the outcomes are shown in table 1.5. From table 1.5 it is confirmed that for a smaller range of a (1–0), search agents gather near the best humpback whale

Table 1.5. Comparison of the WOA-based FODD for different values of n_p, b, and a.

Control parameters		$H_{DD}(z) = \dfrac{a_0 + a_1z^{-1} + a_2z^{-2} + a_3z^{-3} + a_4z^{-4} + a_5z^{-5}}{b_0 + b_1z^{-1} + b_2z^{-2} + b_3z^{-3} + b_4z^{-4} + b_5z^{-5}}$	RMSME (dB)			
			Minimum	Maximum	Mean	SD
Population size n_p	50	$H_{DD}(z) = \dfrac{1.077 - 3.065z^{-1} + 2.865z^{-2} - 0.715z^{-3} - 0.273z^{-4} + 0.112z^{-5}}{1.009 - 2.261z^{-1} + 1.373z^{-2} - 0.115z^{-3} - 0.268z^{-4} + 0.035z^{-5}}$	-40.88	-36.94	-38.49	0.673
	100	$H_{DD}(z) = \dfrac{1.077 - 3.065z^{-1} + 2.865z^{-2} - 0.717z^{-3} - 0.272z^{-4} + 0.112z^{-5}}{1.011 - 2.261z^{-1} + 1.372z^{-2} + 0.114z^{-3} - 0.268z^{-4} + 0.034z^{-5}}$	-43.29	-42.88	-43.21	0.071
	150	$H_{DD}(z) = \dfrac{1.077 - 3.065z^{-1} + 2.864z^{-2} - 0.716z^{-3} - 0.272z^{-4} + 0.112z^{-5}}{1.010 - 2.262z^{-1} + 1.373z^{-2} + 0.113z^{-3} - 0.268z^{-4} + 0.034z^{-5}}$	-43.49	-42.78	-43.01	0.112
Logarithmic spiral shape constant b	0.01	$H_{DD}(z) = \dfrac{1.073 - 3.064z^{-1} + 2.867z^{-2} - 0.718z^{-3} - 0.273z^{-4} + 0.113z^{-5}}{1.012 - 2.261z^{-1} + 1.374z^{-2} + 0.113z^{-3} - 0.268z^{-4} + 0.033z^{-5}}$	-39.95	-39.18	-39.75	0.394
	0.1	$H_{DD}(z) = \dfrac{1.077 - 3.064z^{-1} + 2.864z^{-2} - 0.716z^{-3} - 0.273z^{-4} + 0.112z^{-5}}{1.009 - 2.261z^{-1} + 1.374z^{-2} + 0.116z^{-3} - 0.269z^{-4} + 0.033z^{-5}}$	-42.88	-37.12	-40.85	0.395
	1	$H_{DD}(z) = \dfrac{1.073 - 3.065z^{-1} + 2.864z^{-2} - 0.717z^{-3} - 0.272z^{-4} + 0.112z^{-5}}{1.011 - 2.261z^{-1} + 1.372z^{-2} + 0.114z^{-3} - 0.268z^{-4} + 0.034z^{-5}}$	-43.29	-42.88	-43.21	0.071
	10	$H_{DD}(z) = \dfrac{1.076 - 3.064z^{-1} + 2.866z^{-2} - 0.717z^{-3} - 0.274z^{-4} + 0.114z^{-5}}{1.009 - 2.263z^{-1} + 1.373z^{-2} + 0.112z^{-3} - 0.263z^{-4} + 0.033z^{-5}}$	-37.59	-34.71	-36.48	0.431
Range of a	$\frac{1}{2}$ to 0	$H_{DD}(z) = \dfrac{1.076 - 3.064z^{-1} + 2.866z^{-2} - 0.717z^{-3} - 0.274z^{-4} + 0.115z^{-5}}{1.009 - 2.263z^{-1} + 1.373z^{-2} + 0.112z^{-3} - 0.263z^{-4} + 0.033z^{-5}}$	-40.11	-37.26	-39.15	0.393
	2 to 0	$H_{DD}(z) = \dfrac{1.077 - 3.065z^{-1} + 2.864z^{-2} - 0.717z^{-3} - 0.272z^{-4} + 0.112z^{-5}}{1.011 - 2.261z^{-1} + 1.372z^{-2} + 0.114z^{-3} - 0.268z^{-4} + 0.034z^{-5}}$	-43.29	-42.88	-43.21	0.071
	4 to 0	$H_{DD}(z) = \dfrac{1.076 - 3.064z^{-1} + 2.865z^{-2} - 0.717z^{-3} - 0.273z^{-4} + 0.112z^{-5}}{1.010 - 2.261z^{-1} + 1.373z^{-2} + 0.115z^{-3} - 0.268z^{-4} + 0.032z^{-5}}$	-42.48	-41.77	-41.85	0.154

position, and poor exploration capability is observed. However, by considering a broader range (4–0), the influence due to the best whale's position is reduced, and whales move within the search landscape aimlessly. It results in the deterioration of the exploitation phase as well as solution quality. Considering these two extreme conditions, a rational balance is achieved when the range of 2–0 is considered.

1.3.3 Comparison with existing FODDs

In this subsection, the efficacy of the proposed WOA-based FODD of order 5 for the HOD is compared with all of the state-of-the-art reported FODDs to determine the design efficacy of the proposed FODD, and a comparison is made with the DE [39], PSO [39], RGA [39], GGSA [39], and NMSA [43] algorithm based FODDs. The MR comparison plots of the proposed fifth-order FODD, ideal HOD, and the reported FODDs are shown in figure 1.7(a). The comparison of AME (in dB) for the designed fifth-order FODD for the HOD, along with the reported approaches is illustrated in figure 1.7(b). In figures 1.7(a)–(b), it is shown that the design efficiency

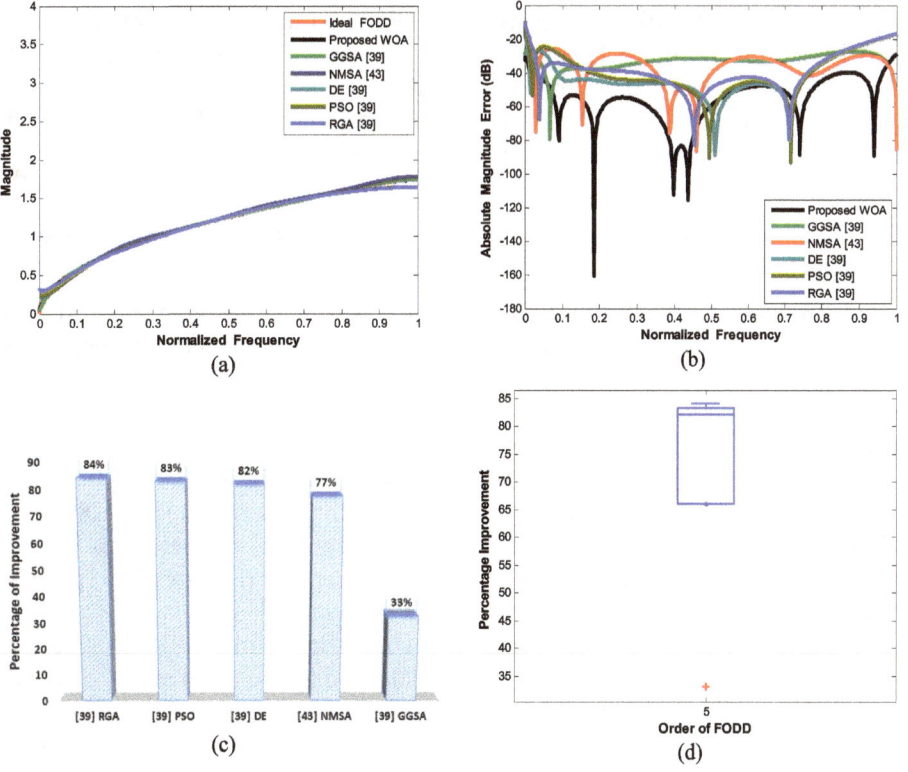

Figure 1.7. Performance comparison of the designed WOA-based FODD of order 5 for the HOD with the published approaches: (a) MR comparison plots, (b) AME comparison plots, (c) percentage improvement of the proposed WOA-based FODD in terms of RMSME in comparison to the reported FODDs, and (d) illustration of average percentage improvement resulted by the WOA-based FODD in comparison with the existing literature.

Table 1.6. Performance comparison of the proposed WOA-based FODD with the reported FODDs.

Literature	Method	RMSME	RMSME (dB)
Mahata *et al* [39]	RGA	0.0426	−27.42
Mahata *et al* [39]	PSO	0.0394	−28.09
Mahata *et al* [39]	DE	0.0373	−28.56
Rana *et al* [43]	NMSA	0.0301	−30.43
Mahata *et al* [39]	GGSA	0.0101	−39.89
Proposed	WOA	0.0068	−43.29

of the proposed WOA-based FODD is improved in terms of magnitude plot and AME plot, respectively. The estimated RMSME of the reported and the proposed fifth-order FODDs are shown in table 1.6.

From table 1.6 it is confirmed that the proposed WOA-based FODD has the least possible RMSME value of 0.0068, whereas Mahata *et al* [39] have reported the best RMSME of 0.0101 for GGSA-based FODD. This ensures an improvement of 33% (= (0.0101–0.0068)/0.0101) in the proposed WOA-based FODD over the NMSA-based FODD [43]. Figure 1.7(c) illustrates the percentage of improvement (POI) resulting in the proposed technique over the existing FODDs. Furthermore, as depicted in figure 1.7(d), an average percentage improvement of 71% is obtained by the proposed technique over all the published FODDs.

1.3.4 Simulation results of the proposed fifth-order FODD for the HOD based QRS detector

The performance of the WOA-FODD based QRS identifier is extensively scrutinized against the AFTDB ECG datasets, and the summary of performance in terms of Se, PP, DER, and Acc metrics is shown in table 1.7. The proposed QRS detector is examined by employing a total of 7592 heartbeats and produces TP, FP, and FN heartbeats of 7542, 15, and 50, respectively. The calculated overall Se, PP, DER, and Acc metrics of the proposed QRS detector are 99.34%, 99.80%, 0.86%, and 99.15%, respectively, whereas the range of each parameter is 93.28%–100%, 96.52%–100%, 0%–5.63%, and 90.24%–100%, respectively. Therefore, it is confirmed that the proposed WOA-FODD based R-peak detector is a promising technique for the identification of the R-peak in the ECG signal.

For clarity, some special cases are also shown in figures 1.8–1.10 where it is observed that the proposed QRS identifier detects the true occurrences of R-peaks successfully even for the cases with some critical conditions: (i) ECG signal with high amplitude T-waves (MITDB record 117) as shown in figures 1.8(a)–(d), (ii) ECG signal affected by severe baseline wandering and high grade of muscle noise (MITDB record 103) as shown in figures 1.10(a)–(d), and (iii) ECG with an irregular heart rate and abrupt peak amplitude variation (MITDB record 228) as shown in figures 1.9(a)–(d). In the MITDB, record 103 is one of the most critical ECG tapes for autonomous QRS peak detection. The high amplitude or hyperacute T-waves

Table 1.7. Summary of the results of the proposed QRS complex detector against AFTDB.

ECG records	Total beats	TP	FP	FN	Se (%)	PP (%)	DER (%)	Acc (%)
LS/n01	76	76	0	0	100.00	100.00	0.00	100.00
LS/n02	73	73	0	0	100.00	100.00	0.00	100.00
LS/n03	50	50	0	0	100.00	100.00	0.00	100.00
LS/n04	92	92	0	0	100.00	100.00	0.00	100.00
LS/n05	96	96	0	0	100.00	100.00	0.00	100.00
LS/n06	69	69	1	0	100.00	98.57	1.45	98.57
LS/n07	71	67	0	4	94.37	100.00	5.63	94.37
LS/n08	87	87	0	0	100.00	100.00	0.00	100.00
LS/n09	75	75	0	0	100.00	100.00	0.00	100.00
LS/n10	89	89	0	0	100.00	100.00	0.00	100.00
LS/s01	159	153	0	6	96.23	100.00	3.77	96.23
LS/s02	107	107	0	0	100.00	100.00	0.00	100.00
LS/s03	62	62	1	0	100.00	98.41	1.61	98.41
LS/s04	121	121	0	0	100.00	100.00	0.00	100.00
LS/s05	67	67	1	0	100.00	98.53	1.49	98.53
LS/s06	70	70	0	0	100.00	100.00	0.00	100.00
LS/s07	80	80	0	0	100.00	100.00	0.00	100.00
LS/s08	103	103	0	0	100.00	100.00	0.00	100.00
LS/s09	82	81	0	1	98.78	100.00	1.22	98.78
LS/s10	116	115	0	1	99.14	100.00	0.86	99.14
LS/t01	154	149	2	5	96.75	98.68	4.55	95.51
LS/t02	110	110	0	0	100.00	100.00	0.00	100.00
LS/t03	57	57	0	0	100.00	100.00	0.00	100.00
LS/t04	126	126	0	0	100.00	100.00	0.00	100.00
LS/t05	67	67	0	0	100.00	100.00	0.00	100.00
LS/t06	70	70	0	0	100.00	100.00	0.00	100.00
LS/t07	105	105	0	0	100.00	100.00	0.00	100.00
LS/t08	107	107	0	0	100.00	100.00	0.00	100.00
LS/t09	87	87	0	0	100.00	100.00	0.00	100.00
LS/t10	126	126	0	0	100.00	100.00	0.00	100.00
TS-a/a01	81	81	0	0	100.00	100.00	0.00	100.00
TS-a/a02	96	94	0	2	97.92	100.00	2.08	97.92
TS-a/a03	98	98	0	0	100.00	100.00	0.00	100.00
TS-a/a04	73	73	0	0	100.00	100.00	0.00	100.00
TS-a/a05	63	63	0	0	100.00	100.00	0.00	100.00
TS-a/a06	78	78	0	0	100.00	100.00	0.00	100.00
TS-a/a07	90	90	0	0	100.00	100.00	0.00	100.00
TS-a/a08	77	77	0	0	100.00	100.00	0.00	100.00
TS-a/a09	73	73	0	0	100.00	100.00	0.00	100.00
TS-a/a10	71	71	0	0	100.00	100.00	0.00	100.00
TS-a/a11	73	73	0	0	100.00	100.00	0.00	100.00
TS-a/a12	62	62	0	0	100.00	100.00	0.00	100.00
TS-a/a13	106	104	0	2	98.11	100.00	1.89	98.11

TS-a/a14	91	89	2	2	97.80	97.80	4.40	95.70
TS-a/a15	117	117	0	0	100.00	100.00	0.00	100.00
TS-a/a16	87	87	0	0	100.00	100.00	0.00	100.00
TS-a/a17	82	82	0	0	100.00	100.00	0.00	100.00
TS-a/a18	72	72	0	0	100.00	100.00	0.00	100.00
TS-a/a19	119	111	4	8	93.28	96.52	10.08	90.24
TS-a/a20	77	77	0	0	100.00	100.00	0.00	100.00
TS-a/a21	83	83	0	0	100.00	100.00	0.00	100.00
TS-a/a22	89	89	0	0	100.00	100.00	0.00	100.00
TS-a/a23	96	96	0	0	100.00	100.00	0.00	100.00
TS-a/a24	115	115	2	0	100.00	98.29	1.74	98.29
TS-a/a25	126	126	0	0	100.00	100.00	0.00	100.00
TS-a/a26	72	72	0	0	100.00	100.00	0.00	100.00
TS-a/a27	85	85	0	0	100.00	100.00	0.00	100.00
TS-a/a28	114	113	0	1	99.12	100.00	0.88	99.12
TS-a/a29	184	181	2	3	98.37	98.91	2.72	97.31
TS-a/a30	93	93	0	0	100.00	100.00	0.00	100.00
TS-b/b01	119	119	0	0	100.00	100.00	0.00	100.00
TS-b/b02	106	106	0	0	100.00	100.00	0.00	100.00
TS-b/b03	124	122	0	2	98.39	100.00	1.61	98.39
TS-b/b04	103	103	0	0	100.00	100.00	0.00	100.00
TS-b/b05	85	85	0	0	100.00	100.00	0.00	100.00
TS-b/b06	65	65	0	0	100.00	100.00	0.00	100.00
TS-b/b07	109	109	0	0	100.00	100.00	0.00	100.00
TS-b/b08	108	106	0	2	98.15	100.00	1.85	98.15
TS-b/b09	122	122	0	0	100.00	100.00	0.00	100.00
TS-b/b10	68	68	0	0	100.00	100.00	0.00	100.00
TS-b/b11	81	81	0	0	100.00	100.00	0.00	100.00
TS-b/b12	148	146	0	2	98.65	100.00	1.35	98.65
TS-b/b13	96	96	0	0	100.00	100.00	0.00	100.00
TS-b/b14	162	158	0	4	97.53	100.00	2.47	97.53
TS-b/b15	73	73	0	0	100.00	100.00	0.00	100.00
TS-b/b16	161	156	0	5	96.89	100.00	3.11	96.89
TS-b/b17	63	63	0	0	100.00	100.00	0.00	100.00
TS-b/b18	77	77	0	0	100.00	100.00	0.00	100.00
TS-b/b19	83	83	0	0	100.00	100.00	0.00	100.00
TS-b/b20	142	142	0	0	100.00	100.00	0.00	100.00
Total	7592	7542	15	50	99.34	99.80	0.86	99.15

LS: Learning-Set, TS: Test-Set.

can be an early sign of ST-elevation myocardial infarction. Similarly, the irregular heart rate, as shown in figure 1.9(a) needs immediate clinical attention. In figures 1.8–1.10 it is clear that all the R-peaks are identified successfully by the proposed technique. It is because of the robust feature-inducing ability of the proposed FODD based feature signal-generating stage.

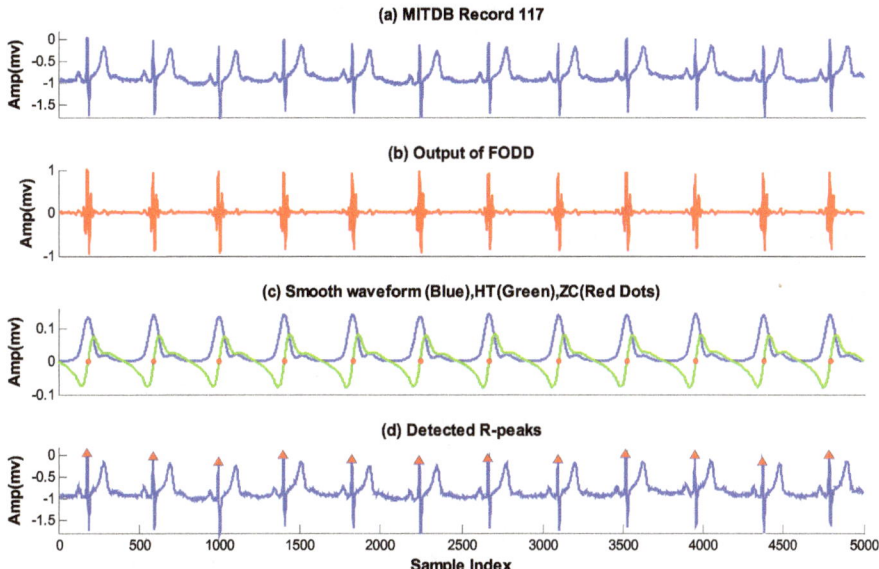

Figure 1.8. QRS detection of ECG signal with elevated T-waves (record 117 of the MITDB): (a) raw ECG signal, (b) FODD output, (c) smooth waveform (blue), Hilbert transform output (green), and zero crossing (red dots), and (d) detected R-peaks.

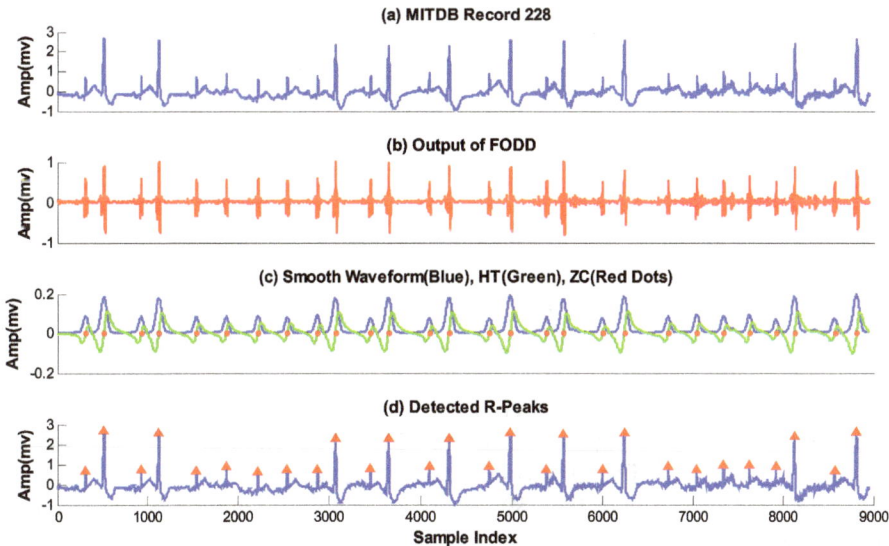

Figure 1.9. QRS detection of ECG signal with an abrupt change in QRS complex amplitude (record 228 of the MITDB): (a) raw ECG signal, (b) FODD output, (c) smooth waveform (blue), Hilbert transform output (green), and zero crossing (red dots), and (d) detected R-peaks.

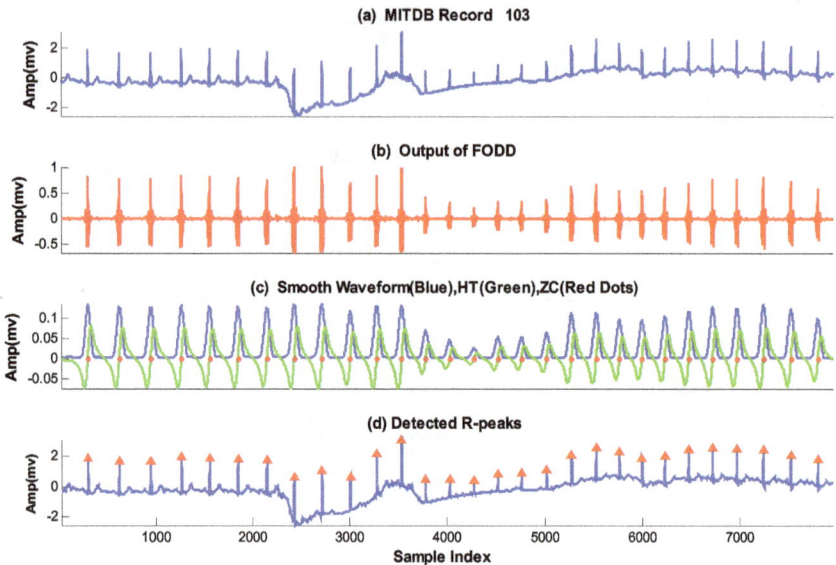

Figure 1.10. QRS detection of ECG signal with baseline shift (record 103 of the MITDB): (a) raw ECG signal, (b) FODD output, (c) smooth waveform (blue), Hilbert transform output (green), and zero crossing (red dots), and (d) detected R-peaks.

Table 1.8. Performance comparison with the previous studies.

Publication	Total beats	TP beats	FP beats	FN beats	Se (%)	PP (%)	DER (%)	Acc (%)
Elgendi [27]	7618	N/R	N/R	N/R	99.7	99.7	*	*
Burgera [28]	7618	N/R	N/R	N/R	88.6	96.9	*	*
Proposed method	7592	7542	15	50	99.34	99.80	0.86	99.15

1.3.4.1 Performance comparison

This subsection compares the performance indices of the proposed R-peak detection methodology with the performance indices of the previous studies to express the proficiency of the proposed approach. All of those QRS detectors are evaluated on the first channel (ECG) tapes of the AFTDB. The comparative outcomes are presented in table 1.8. The proposed R-peak detection methodology accurately detects 7542 *TP* beats out of the 7592 beats. This confirms that the proposed QRS identification methodology is efficient in accurately detecting 99.15% (= 7542/(7542 + 15 + 50)) of the total number of heartbeats. Compared to the published advanced QRS detection methods [27, 28], a superior WOA-based fractional-order differentiator operated QRS complex detector is proposed here, and either better or comparative performance has been confirmed in terms of the Se, PP, and Acc metrics.

1.4 Conclusion

The IIR fractional-order digital differentiator (FODD), which is employed in this article for reliable and accurate ECG R-peak identification, is implemented using the whale optimization algorithm (WOA). With the aid of the WOA's carefully calibrated control settings, it is capable of arriving at the lowest error cost value in the fewest iterations. As a result of this, a highly accurate magnitude characteristic of the suggested FODD is achieved, and it outperforms the conventional FODD design approaches. The proposed WOA-based FODD has now been found to significantly increase the performance of the QRS detector when it serves as a crucial component of the pre-processor which yields high sensitivity (99.34%), high positive predictivity (99.80%), and high accuracy (99.15%) with a little detection error rate of 0.86%. This makes the proposed FODD based QRS complex detector a viable method for the identification of R-peak in the ECG wave.

References

[1] World Health Organization 2013 *World Health Statistics* (Geneva: World Health Organization)

[2] Nayak C, Saha S K, Kar R and Mandal D 2020 Efficient design of zero-phase Riesz fractional order digital differentiator using manta-ray foraging optimization for precise electrocardiogram QRS detection *IEEE Open J. Circuits Syst.* **1** 280–92

[3] Berkaya S K, Uysal A K, Gunal E S, Ergin S, Gunal S and Gulmezoglu M B 2018 A survey on ECG analysis *Biomed. Signal Process. Control* **43** 216–35

[4] Nuryani N, Ling S S H and Nguyen H T 2012 Electrocardiographic signals and swarm-based support vector machine for hypoglycemia detection *Ann. Biomed. Eng.* **40** 934–45

[5] Kohler B U, Hennig C and Orglmeister R 2002 The principles of software QRS detection *IEEE Eng. Med. Biol.* **21** 42–57

[6] Min Y J, Kim H K, Kang Y R, Kim G S, Park J and Kim S W 2013 Design of wavelet-based ECG detector for implantable cardiac pacemakers *IEEE Trans. Biomed. Circuits Syst.* **7** 426–36

[7] Benmalek M and Charef A 2009 Digital fractional-order operators for R wave detection in the electrocardiogram signal *IET Signal Proc.* **3** 381–91

[8] Pan J and Tompkins W J 1985 A real-time QRS detection algorithm *IEEE Trans. Biomed. Eng.* **32** 230–6

[9] Nayak C, Saha S K, Kar R and Mandal D 2018 An efficient QRS complex detection using optimally designed digital differentiator *Circuits Syst. Signal Process.* **38** 716–49

[10] Yakut O and Bolat E D 2018 An improved QRS complex detection method having a low computational load *Biomed. Signal Process. Control* **42** 230–41

[11] Manikandan M S and Soman K P 2012 A novel method for detecting R-peaks in electrocardiogram (ECG) signal *Biomed. Signal Process. Control* **7** 118–28

[12] Nayak C, Saha S K, Kar R and Mandal D 2018 Automated QRS complex detection using MFO-based DFOD *IET Signal Proc.* **12** 1172–84

[13] Arzeno N M, Deng Z D and Poon C S 2008 Analysis of first-derivative based QRS detection algorithms *IEEE Trans. Biomed. Eng.* **55** 478–84

[14] Ferdi Y, Herbeuval J P, Charef A and Boucheham B 2000 R wave detection using fractional digital differentiation *ITBM-RBM* **24** 273–80

[15] Thakor N V, Webster J G and Tompkins W J 1983 Optimal QRS detector *Med. Biol. Eng. Comput.* **21** 343–250

[16] Nallathambi G and Príncipe J C 2014 Integrate and fire pulse train automaton for QRS detection *IEEE Trans. Biomed. Eng.* **61** 317–26

[17] Deepu C J and Lian Y 2015 A joint QRS detection and data compression scheme for wearable sensors *IEEE Trans. Biomed. Eng.* **62** 165–75

[18] Fraden J and Neuman M R 1980 QRS wave detection *Med. Biol. Eng. Comput.* **18** 125–32

[19] Dohare A K, Kumar V and Kumar R 2014 An efficient new method for the detection of QRS in the electrocardiogram *Comput. Electr. Eng.* **40** 1717–30

[20] Phukpattaranont P 2015 QRS detection algorithm based on the quadratic filter *Expert Syst. Appl.* **42** 4867–77

[21] Castells-Rufas D and Carrabina J 2015 Simple and real-time QRS detector with the MaMeMi filter *Biomed. Signal Process. Control* **21** 137–45

[22] Arbateni K and Bennia A 2014 Sigmoidal radial basis function ANN for QRS complex detection *Neurocomputing* **145** 438–50

[23] Hou Z, Dong Y, Xiang J, Li X and Yang B 2018 A real-time QRS detection method based on phase portraits and box-scoring calculation *IEEE Sens. J.* **18** 3694–702

[24] Tang X, Hu Q and Tang W 2018 A real-time QRS detection system with PR/RT interval and ST-segment measurements for wearable ECG sensors using parallel delta modulators *IEEE Trans. Biomed. Circuits Syst.* **12** 751–61

[25] Yazdani S, Fallet S and Vesin J M 2018 A novel short-term event extraction algorithm for biomedical signals *IEEE Trans. Biomed. Eng.* **65** 754–62

[26] Lee J W, Kim K S, Lee B and Lee M H 2002 A real-time QRS detection using delay-coordinate mapping for the microcontroller implementation *Ann. Biomed. Eng.* **30** 1140–51

[27] Elgendi M 2013 Fast QRS detection with an optimised knowledge-based method: evaluation on 11 standard ECG databases *PLoS One* **8** 1–18

[28] Burguera A 2018 Fast QRS detection and ECG compression based on signal structural analysis *IEEE J. Biomed. Health Inform.* **23** 123–31

[29] Chen B, Huang S, Lian Z, Chen W and Pan B 2018 A fractional-order derivative-based active contour model for inhomogeneous image segmentation *Appl. Math. Modell.* **65** 120–36

[30] Pu Y F, Zhou J L and Yuan X 2010 Fractional differential mask: a fractional differential-based approach for multiscale texture enhancement *IEEE Trans. Image Process.* **19** 491–511

[31] Victor S, Melchior P, Pellet M and Oustaloup A 2020 Lung thermal transfer system identification with fractional models *IEEE Trans. Control Syst. Technol.* **28** 172–82

[32] Mescia L, Bia P and Caratelli D 2014 Fractional derivative-based FDTD modelling of transient wave propagation in Havriliak–Negami media *IEEE Trans. Microwave Theory Tech.* **62** 1920–9

[33] Nayak C, Saha S K, Kar R and Mandal D 2021 Optimal design of zero-phase digital Riesz FIR fractional order differentiator *Soft Comput.* **25** 4261–82

[34] Chareff A, Sun H H, Tsao Y Y and Onaral B 1992 Fractal system as represented by singularity function *IEEE Trans. Autom. Control* **37** 1465–70

[35] Barbosa R S, Machado J A T and Silva M F 2006 Time domain design of fractional differintegrators using least-squares *Signal Process.* **86** 2567–81

[36] Tolba M F, Said L A, Madian A H and Radwan A G 2018 The FPGA implementation of the fractional order integrator/differentiator: two approaches and applications *IEEE Trans. Circuits Syst.* I **66** 1484–95

[37] Vinagre B M, Chen Y Q and Petras I 2003 Two direct Tustin discretization methods for fractional order differentiator/integrator *J. Franklin Inst.* **340** 349–62

[38] Al-Moalmi A, Luo J, Salah A, Li K and Yin L 2021 A whale optimization system for energy-efficient container placement in data centers *Expert Syst. Appl.* **164** 113719

[39] Mahata S, Saha S K, Kar R and Mandal D 2016 Optimal and accurate design of fractional order digital differentiator—an evolutionary approach *IET Signal Proc.* **11** 181–96

[40] Nayak C, Saha S K, Kar R and Mandal D 2019 An efficient and robust digital fractional order differentiator based ECG pre-processor design for QRS detection *IEEE Trans. Biomed. Circuits Syst.* **13** 682–96

[41] Nayak C, Saha S K, Kar R and Mandal D 2019 Optimal SSA-based wideband digital differentiator design for cardiac QRS complex detection application *Int. J. Numer. Model.: Electron. Netw., Devices Fields* **32** e2524

[42] Pham Q V, Mirjalili S, Kumar N, Alazab M and Hwang W J 2020 Whale optimization algorithm with applications to resource allocation in wirless networks *IEEE Transactions on Vehicular Technology* **69** 4285–97

[43] Rana K P S, Kumar V, Garg Y and Nair S S 2016 Efficient design of discrete fractional-order differentiators using Nelder–Mead simplex algorithm *Circuits Syst. Signal Process.* **35** 2155–88

[44] Nayak C, Saha S K, Kar R and Mandal D 2020 Full band IIR digital differentiators design using evolutionary algorithm Proc. IEEE Advanced Communication Technologies and Signal Processing (ACTS) *(Silchar, India, 4–6 Dec)* pp 1–6

[45] Nayak C, Saha S K, Kar R and Mandal D 2021 Efficient design of DFODs using GBMO Proc. IEEE 6th Int. Conf. for Convergence in Technology (I2CT) *(Maharashtra, India, 2–4 April)* pp 1–6

[46] Mirjalili S and Lewis A 2016 The whale optimization algorithm *Adv. Eng. Software* **95** 51–67

[47] Hussain K, Xia Y, Onaizah A N, Manzoor T and Jalil K 2022 Hybrid of WOA-ABC proposed CNN for intrusion detection system in wireless sensor networks *Optik* **271** 170145

[48] Singh A and Khamparia A 2020 Hybrid whale optimization-differential evolution and genetic algorithm based approach to solve unit commitment scheduling problem *WODEGA, Sustain. Comput.: Inform. Syst.* **28** 100442

[49] Nayak C, Saha S K, Kar R and Mandal D 2019 An optimally designed digital differentiator based pre-processor for R-peak detection in electrocardiogram signal *Biomed. Signal Process. Control* **49** 440–64

[50] Goldberger A L, Amaral L A N, Glass L, Hausdorff J M, Ivanov P C, Mark R G, Mietus J E, Moody G B, Peng C K and Stanley H E 2000 PhysioBank, PhysioToolkit, and PhysioNet components of a new research resource for complex physiologic signals *Circulation* **101** e215–20

[51] Moody G B and Mark R G 2001 The impact of MIT-BIH arrhythmia database *IEEE Eng. Med. Biol. Mag.* **20** 45–50

[52] Moody G B 2004 Spontaneous termination of atrial fibrillation: a challenge from PhysioNet and computers in cardiology *Comput. Cardiol.* **31** 101–4

Chapter 2

Focal and non-focal EEG signal classification using the Wigner–Ville distribution and deep feature extraction

Sachin Taran, Smith K Khare, P V Keshava Krishna and Vikram Singh Kardam

Accurate detection of focal epilepsy is helpful in locating the epileptogenic region for effective surgery. Electroencephalogram (EEG) signals capture the changes in the brain during focal epilepsy. However, capturing representative information from EEG signals is difficult because of their dynamically changing nature. Therefore, this chapter used a novel hybrid model for accurate and effective detection of focal EEG signals. The methodology is composed of four stages. In the first stage, the Wigner–Ville distribution converts focal and non-focal EEG signals to a time–frequency distribution. The second stage is deep feature extraction based on the pre-trained AlexNet CNN model. In the third stage, the deep feature classification is executed using an extreme learning machine classifier. Finally, the system's performance is evaluated using different performance models in the last stage. Our developed hybrid model has yielded the accuracy, sensitivity, and specificity of 98.94%, 98.72%, and 99.17%, respectively.

2.1 Introduction

The non-invasive mode of recording an electroencephalogram (EEG) is effective for the detection of different neurological diseases such as epilepsy, Parkinson's disease, attention deficit hyperactivity disorder, and schizophrenia [1–4]. Focal epilepsy is the term used to describe epileptic seizures that only impact a specific brain area. Parts of the brain are affected by partial epilepsy, and the EEG taken from those areas is called focal (F) EEG, while the EEG taken from other areas is called non-focal (NF) EEG [5]. Identifying the focal EEG signal offers a method to locate the epileptogenic region for effective surgery [6]. Because EEG is accurate and can measure brain activity with a resolution of milliseconds, it is a fast and safe method for observing brain activity [7].

Numerous investigations have been carried out to detect F EEG signals. The clustering variational mode decomposition with an extreme learning machine (ELM) classifier has been explored [8]. Tunable Q-factor wavelet transform (TQWT) based features were investigated to classify F and NF EEG data using neural network (NN) classifiers [9]. Several entropy features were extracted from the Bern-Barcelona database and given to multiple classifiers for the classification of F and NF EEG signals [5]. The computerized automated detection of focal epileptic seizures was introduced for the classification of F and NF EEG signals using neighbourhood component analysis-based feature extraction and k nearest neighbour (KNN), support vector machine (SVM), and random forest classifiers [10]. Empirical mode decomposition (EMD) was employed as a feature extraction method, and the generated features were given to the least squares SVM classifier [11]. The synchro-squeezing transform with a deep convolutional NN was introduced to automatically categorize F and NF EEG signals [12]. EMD was used to extract the log energy entropy features, which were then fed into the KNN and SVM classifiers for classification [13]. The least-square SVM classifier fed with empirical wavelet transform (WT)-based EEG rhythms was used to automatically identify F and NF EEG signals [14]. Using a softmax classifier, deep NN-based categorization of F and NF EEG data was examined in [15].

A combination of VMD and discrete WT-based features was been explored. The ensemble stacking classification model was introduced for the classification of F and NF EEG data in [16]. The least squares SVM classifier was used to explore discrete Fourier transform-based EEG rhythms for the classification of F and NF EEG signals [16]. An SVM classifier was used to explore multi-features taken from multiple domains to classify F and NF EEG data [17]. The tool for feature extraction was autoregressive. To classify F and NF epileptic EEG signals, the collected features were further compared using various machine-learning methods [18]. Using a long-short-term memory technique, F and NF epileptic EEG signals can be identified automatically [19]. A dual-tree complex WT with an adaptive neuro-fuzzy inference system classifier was used to explore the computer-aided automatic detection of F EEG signals [20]. For the categorization of F and NF EEG signals, the Hilbert–Huang transform-based EEG rhythms were extracted using EMD and supplied to the least squares SVM [21]. A novel approach called the F and NF index was introduced for identifying F and NF EEG signals using the discrete WT and several entropy features [22]. The automatic classification of F and NF EEG signals using a unique method that combines NN and EMD has been proposed [23]. To analyse and classify F and NF EEG signals, the complexity of multivariate EEG signals at various frequency scales was examined using TQWT [24]. F and NF EEG signals were classified using the variance and entropies extracted using EMD [25]. The log energy entropy (LEE) feature was investigated using EMD and SVM to classify F and NF EEG signals [26]. Automatic identification of F and NF EEG signals was introduced with TQWT and LS-SVM [27]. The FAWT-based features were investigated using KNN and SVM classifiers to identify F EEG signals [28].

The literature shows that pre-defined basis functions limit existing models. Due to the non-linear nature of EEG signals, pre-defined basis functions may not yield the desired performance. In addition, signal analysis in only one domain can result in performance degradation. Therefore, this model explores the analysis of time–frequency characteristics of EEG signals simultaneously using the Wigner–Ville distribution. The deep features are extracted using a pre-trained AlexNet convolutional neural network model and classified using an extreme learning machine classifier. We have evaluated the efficacy of our developed model using a ten-fold cross-validation technique and different performance measures.

2.2 Methodology

A block diagram of the proposed work is shown in figure 2.1. The acquisition, OAS, time–frequency distribution, decision-making, and deep features are described in the following subsections.

2.2.1 Dataset

The online available dataset www.dtic.upf.edu/ralph/sc/ was used in this work [6]. Figure 2.2 shows the NF and F EEG signals. A publicly available database of people with drug-resistant temporal lobe epilepsy serves as the confirmed dataset, highlighting the extracted variables' extensive statistical relevance [8]. Five epilepsy patients who were admitted for surgery are included in the collection of recordings. The patients' EEG signals are at random, captured on adjacent channel (x and y) pairs. The 512 Hz sample rate was used to record the 20 s EEG waves. Each F and NF class in this study has 750 EEG signals.

2.2.2 Optimum allocation sampling (OAS)

The term 'optimal allocation' refers to a sample allocation method combined with stratified sampling. At the lowest computational cost, OAS splits lengthy

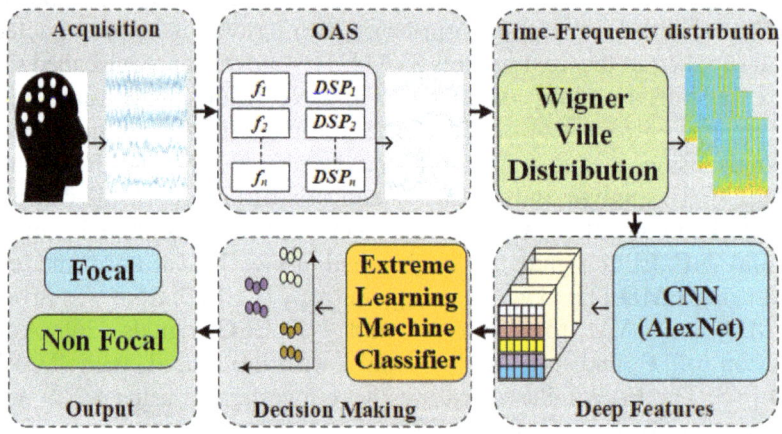

Figure 2.1. Block diagram of the proposed work.

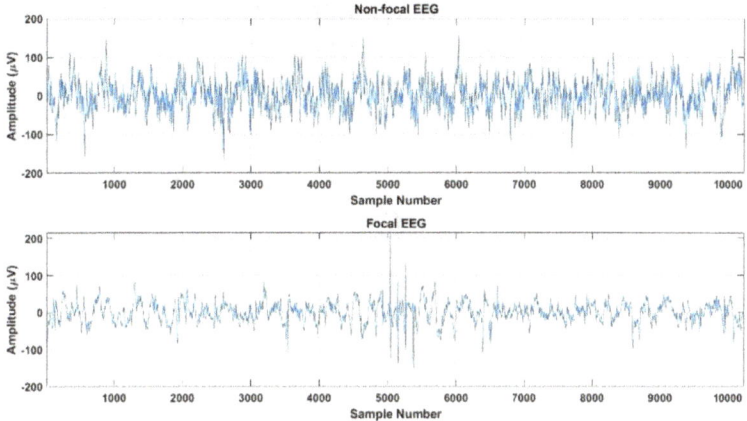

Figure 2.2. The F and NF EEG signals.

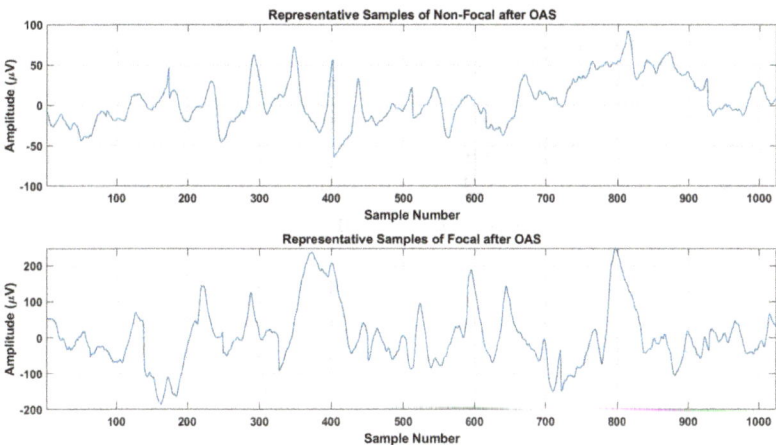

Figure 2.3. Samples of F and NF after optimum allocation sampling (OAS).

non-cognate signals into related groups and determines the pattern points from each group [9, 10]. OAS can handle large non-homogeneous sequence sizes with the least computing expense. The lengthy, non-homogeneous sequence is divided into a few homogeneous groups using the OAS technique, which then chooses the representative sample points from each group. In a broken environment, the sound sequence is divided into stationary groups or subsets, and the most advantageous pattern is selected with the least amount of justification. Instead of calculating the stationary and related ambient sound categorization time array, the non-linear must be evaluated. Figure 2.3 shows the optimum allocation sampling for F and NF EEG signals. From this figure, it can be noted that the samples of the F and NF EEG epoch after OAS shows subtle changes when the time duration of 1000 sample points is taken.

By removing unnecessary data, feature extraction aims to acquire some useful values (data) known as features. It creates a feature vector out of the initial signals. By utilizing a sample size calculator with the necessary interval and assurance level and the length of a large characteristic, it is possible to calculate the initial overall pattern size n. Following the equations that take into account the instability in each group, the OAS method is then used to calculate the pattern size for each group [8, 11, 12, 29]:

$$a(r) = \frac{A_r \sqrt{\sum_{s=1}^{p} W_{rs}^2}}{\sum_{r=1}^{t} N_r \sqrt{\sum_{s=1}^{p} W_{rs}^2}} \times a \tag{2.1}$$

$$r = 1, 2, \ldots, k, \text{ and } s = 1, 2, \ldots, p'$$

$A_1, A_2, \ldots, A_t$ is the total of all achieved sample sizes equal to a, i.e. $a(1) + a(2) + \ldots + a(t) = a$, if $a(1), a(2), \ldots, a(t)$ indicate the sample size obtained.

2.2.3 Wigner–Ville distribution

The WVD is currently being utilized to classify a variety of biological signals [13, 14]. It is a novel transform that makes it possible to analyse time–frequency signals with high-125 resolution. It has been widely used in signal estimation, visualization, and detection [15, 16]. Figure 2.4 shows the time–frequency distribution (TFD) for both F and NF EEG signals. It can be observing from the figure 2.4 that the TFD is different for both F and NF signals.

The classification of CAP phases for distinct frequency ranges was done using EEG data. As a result, it was discovered that time–frequency analysis methods were essential for understanding these signals. Typically, for the given discrete signal $x(n)$ with N samples, the WVD is determined using the following methods:

$$X(m, k) = \sum_{n=-N}^{N} \left(x\left(m + \frac{n}{2}\right) x^*\left(m - \frac{n}{2}\right) e^{\frac{-j2\pi kn}{N}} \right). \tag{2.2}$$

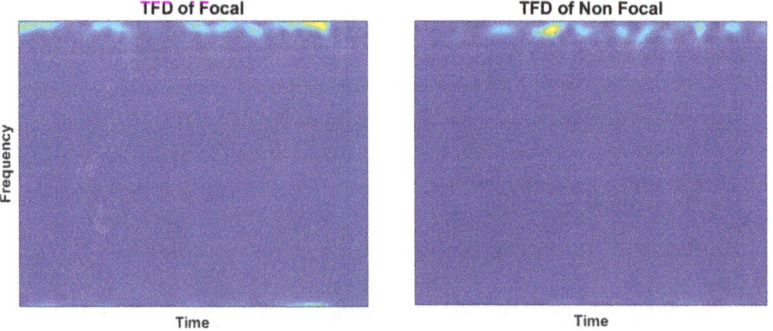

Figure 2.4. Time frequency distribution (TFD) plots for F and NF EEG signals.

Here, $j = \sqrt{-1}$ and m is time and k is frequency components. The 2 s EEG signal is represented here by $x(n)$. The resulting WVD 2D matrix is $N_f \times 2N$, where N_f is the total frequency components. The highest frequency value for sample frequencies of 512 Hz is 256 Hz. The 121 total frequency components of the 135 relevant frequencies for the taken-into-account EEG signals have a frequency range of 0–30 Hz. As a result, the WVD matrix has dimensions: 121×2048 [17].

2.2.4 Deep learning model

This work uses the deep learning algorithm AlexNet. Deep learning was first described in an article that was published in 1998 [18]. However, after 2012, deep learning gained popularity as a result of AlexNet model's victory in the ImageNet competition [19–21].

The most difficult ImageNet visual object recognition task, the ImageNet Large Scale Visual Detection Challenge (ILSVRC), was won in 2012 by Alex Krizhevesky and others who suggested a deeper and wider CNN model than LeNet [23]. For problems involving visual recognition and classification, there was a significant advancement in computer vision and machine learning, which rapidly increased interest in deep learning. Figure 2.5 shows the AlexNet architecture. The first convolutional layer conducts convolution, max-pooling, and local response normalization using the 96 unique 11×11-sized receptive filters (LRN) [30]. The stride size of the 3×3 filters utilized in maximum pooling operations is 2. The 5×5 filters in the second layer accomplish the same thing. The 3×3 filters are used in the third, fourth, and fifth convolutional layers, which use 384, 384, and 296 feature maps, respectively [31]. After dropout and before a softmax layer two fully connected (FC) layers are used. Two parallel networks are trained for this model, both of which have an equal number of feature maps and a comparable structure.

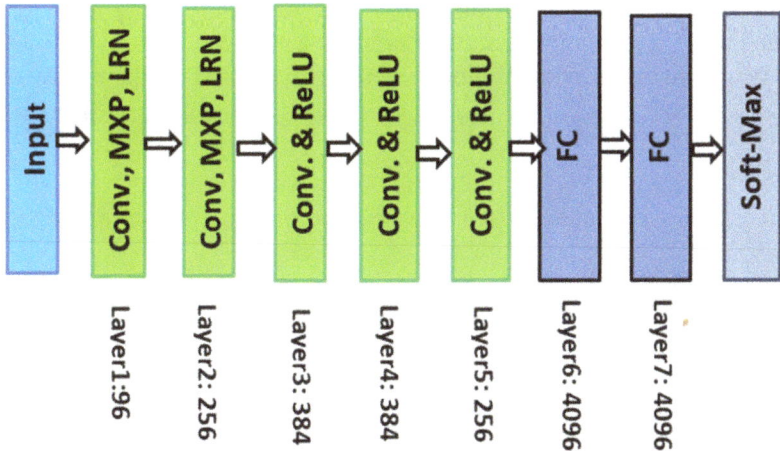

Figure 2.5. Architecture of AlexNet: convolution, max-pooling, LRN, and FC layer.

Two unique ideas—dropout and local response normalization (LRN)—are presented in this network. An NN patch from the same feature map is chosen and normalized using the neighbourhood values when LRN is applied to single channels or feature maps. Additionally, LRN may include channels or maps.

Three convolutional layers and two fully connected layers comprise AlexNet. When using the ImageNet dataset, the total number of parameters for AlexNet can be estimated as follows: the input samples are $224 \times 224 \times 3$, the filters (kernels or masks) or receptive field have a size of 11, the stride is 4, and the first convolution layer's output is $55 \times 55 \times 96$ [19].

2.2.5 ELM classifier

The ELM is a single hidden-layer feed-forward neural network [29, 32–35]. ELM was created to manage data sets with simple architecture and high levels of dimension [36]. Because of its improved generalization performance and quick training, it is appropriate for multi-class classification [29]. The weighted matrix's random construction makes it possible to immediately assess the output weights for classification. Improper learning rates, over-fitting, local minima, and other issues with conventional gradient-based techniques are all fixed by ELM. ELM can also be employed with non-linear activation functions [32, 33, 36, 37].

A single hidden-layer feed-forward network's activation function, $f(x)$, for N numbers of samples chosen at random (x_j, t_j) can be defined as [36, 37]

$$\sum_{i=1}^{N} \beta_i f \left(w_i . \, x_j + b_i \right) = o_j \qquad (2.3)$$

$$j = 1, 2, \dots, N,$$

where a weight vector between input neurons with the ith hidden neuron $w_i = [w_{i1}, w_{i2}, w_{i3}, \dots w_{in}]^T$ and a weight vector $\beta_i = [\beta_{i1}, \beta_{i2}, \beta_{i3}, \dots, \beta_{im}]^T$ connects the output node with ith hidden node, b_i is the threshold, whose inner product is represented by $w_i . \, x_j$. To approximate N samples with zero error, a single-layer feed-forward neural network with N hidden neurons was used in the above equation, which indicates that $\sum_{i=1}^{H} \|o_i - t_j\| = 0$. Now, the above equation can be written as [37]

$$\sum_{i=1}^{N} \beta_i f \left(w_i . \, x_j + b_i \right) = t_j \qquad (2.4)$$

$$j = 1, 2, \dots, N.$$

The above N equations can be redefined as [34, 37, 38]

$$H\beta = T, \qquad (2.5)$$

where H is the hidden layer's output vector and the output weight vector (β) is created by applying the least squares (LS) method to solve the previous equation [34, 37, 38]:

$$L\left(X,\,Y;\hat{\beta}\right) = \frac{\min}{\beta}\,\|\,H\,\beta - Y\,\|^2.$$ (2.6)

The unique LS solution $\hat{\beta}$ can be given as

$$\hat{\beta} = H^\dagger\,Y,$$ (2.7)

where $H^\dagger$ is termed as the Moore–Penrose generalized vector, the inverse of the H vector [34, 38].

2.3 Results and discussion

The deep learning model AlexNet is examined to classify a new class of objects [39, 40]. Additionally, ten-fold cross-validation is used to measure the performance of the network [41, 42]. In the experiment, 10% of the data are used for testing, and 90% of the data are used for training to measure the model's performance. In table 2.1 the classification results obtained using AlexNet FC6 and similarly in table 2.2 the classification results obtained using AlexNet FC7 of EEG data are given.

The evaluation of the proposed model is performed using the performance parameters. Table 2.3 shows the classification accuracy scores for all deep CNNs models. For the AlexNet model, the FC6 and FC7 layers are used. The performance parameters obtained for AN FC6 are as follows: accuracy is 95.72 ± 0.8, sensitivity is 95.22 ± 0.887, specificity is 96.21 ± 0.902, precision is 96.18 ± 0.89, the F1-score is 95.7 ± 0.801, and Matthews correlation coefficient is 91.45 ± 1.598. The performance parameters obtained for AN FC7 are as follows: accuracy is 98.94 ± 0.223, sensitivity is 98.72 ± 0.354, specificity is 99.17 ± 0.281, precision is 99.17 ± 0.28, the F1-score is 98.94 ± 0.224, and Matthews correlation coefficient is 97.89 ± 0.445.

It can be observed from the table 2.1 that 95.72% and 98.94% accuracy scores are provided by the AlexNet FC6 layer and AlexNet FC7 layer, respectively. It is worth

Table 2.1. The classification results for AlexNet FC6 layer.

FC6	ACC	SEN	SPC	F1	MCC	PRC
Fold-1	94.53	94.40	94.67	94.53	89.07	94.65
Fold-2	97.47	97.33	97.60	97.46	94.93	97.59
Fold-3	96.13	94.93	97.33	96.09	92.29	97.27
Fold-4	95.60	95.20	96.00	95.58	91.20	95.97
Fold-5	96.80	95.47	98.13	96.76	93.63	98.08
Fold-6	94.00	94.67	93.33	94.04	88.01	93.42
Fold-7	95.07	94.67	95.47	95.05	90.14	95.43
Fold-8	94.27	92.80	95.73	94.18	88.57	95.60
Fold-9	98.00	98.13	97.87	98.00	96.00	97.87
Fold-10	95.33	94.67	96.00	95.30	90.67	95.95
Average %	**95.72**	**95.23**	**96.21**	**95.70**	**91.45**	**96.18**

Table 2.2. The classification results for the AlexNet FC7 layer.

FC7	ACC	SEN	SPC	F1	MCC	PRC
Fold-1	98.67	98.67	98.67	98.67	97.33	98.67
Fold-2	98.40	97.33	99.47	98.38	96.82	99.46
Fold-3	99.33	98.93	99.73	99.33	98.67	99.73
Fold-4	99.07	98.40	99.73	99.06	98.14	99.73
Fold-5	99.33	99.47	99.20	99.33	98.67	99.20
Fold-6	98.80	98.67	98.93	98.80	97.60	98.93
Fold-7	98.53	98.67	98.40	98.54	97.07	98.40
Fold-8	99.20	98.93	99.47	99.20	98.40	99.46
Fold-9	98.67	98.67	98.67	98.67	97.33	98.67
Fold-10	99.47	99.47	99.47	99.47	98.93	99.47
Average %	**98.95**	**98.72**	**99.17**	**98.94**	**97.90**	**99.17**

Table 2.3. The prediction results for the deep models with the ELM classifier.

Performance parameters

	AN FC6	Error	AN FC7	Error
ACC	95.72	±0.8	98.94	±0.223
SEN	95.22	±0.887	98.72	±0.354
SPC	96.21	±0.902	99.17	±0.281
F1	95.7	±0.801	98.94	±0.224
MCC	91.45	±1.598	97.89	±0.445
PRC	96.18	±0.89	99.17	±0.28

mentioning that the FC7 layers of the AlexNet provide better results than the FC6 layers.

Figures 2.6 and 2.7 represent the hidden node behaviour for AN FC6 and AN FC7 of eleven hidden neurons (110, 120, 130, 140, 150, 160, 170, 180, 190, 200, 210) with respect to four kernels, namely SIG, SINE, TB, and RBF, respectively. It can be observed from figures 2.6 and 2.7 that the SIG kernel performed better F and NF classification among all the four deep network kernels. It can be observed that the SIG kernel obtained the maximum for AN FC6 and AN FC7 models. The best accuracy achieved by AN FC6 is 95.72% with 190 and 210 hidden neurons, whereas the best accuracy achieved by AN FC7 is 98.94% with 210 hidden neurons. No further neurons have been taken because the results clearly show that after 200 to 210 neurons, accuracy is going to be constant.

In figure 2.6 the TB kernel started with very low accuracy. We observed a consistent trend of up to 200 nodes and the accuracy remained constant for 200 and 210 nodes. Similarly, RBF increases persistently from 100 up to 210 without fluctuations. For SINE, initially, the trend is a decrease in accuracy for the 85 to

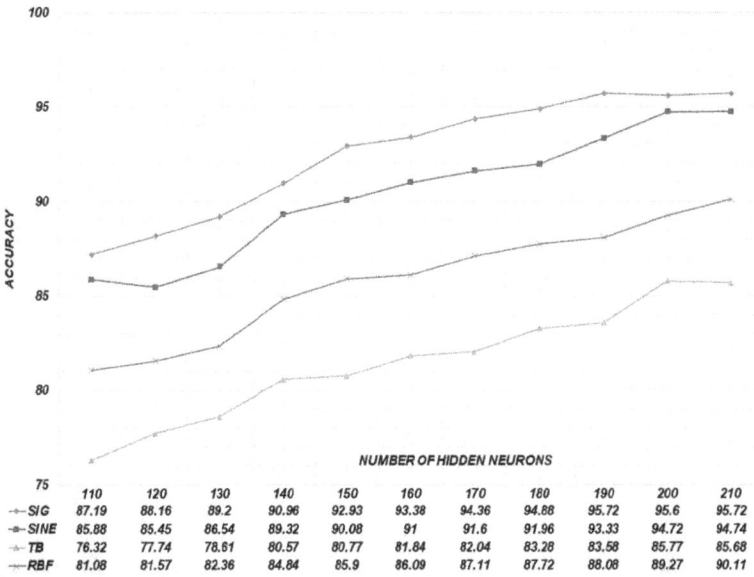

	110	120	130	140	150	160	170	180	190	200	210
SIG	87.19	88.16	89.2	90.96	92.93	93.38	94.36	94.88	95.72	95.6	95.72
SINE	85.88	85.45	86.54	89.32	90.08	91	91.6	91.96	93.33	94.72	94.74
TB	76.32	77.74	78.61	80.57	80.77	81.84	82.04	83.28	83.58	85.77	85.68
RBF	81.08	81.57	82.36	84.84	85.9	86.09	87.11	87.72	88.08	89.27	90.11

Figure 2.6. The classification accuracy plot of AN FC6 for different kernels.

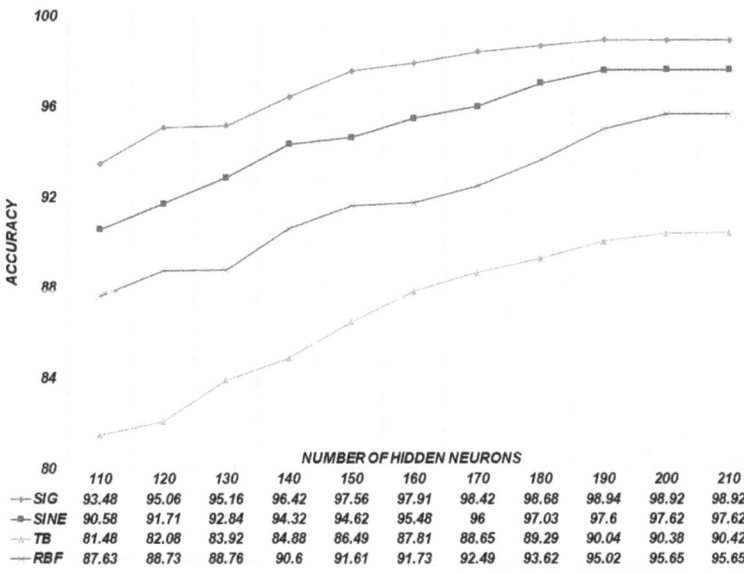

	110	120	130	140	150	160	170	180	190	200	210
SIG	93.48	95.06	95.16	96.42	97.56	97.91	98.42	98.68	98.94	98.92	98.92
SINE	90.58	91.71	92.84	94.32	94.62	95.48	96	97.03	97.6	97.62	97.62
TB	81.48	82.08	83.92	84.88	86.49	87.81	88.65	89.29	90.04	90.38	90.42
RBF	87.63	88.73	88.76	90.6	91.61	91.73	92.49	93.62	95.02	95.65	95.65

Figure 2.7. The classification accuracy plot of AN FC7 for different kernels.

100 nodes, followed by an abrupt increase in accuracy up to 200 nodes and it then remained constant for 200 and 210 nodes. However, SIG showed a persistent increase from 100 to 190, a slight decreased to 200 nodes, and then remained stable for the next 200 and 210 nodes.

Table 2.4. Performance comparison of existing works using the same dataset.

Authors	Proposed method and classifier	Features used	Accuracy (%)
Sharma *et al* [22]	EMD and LS-SVM	Entropy	84
Bhattacharyya *et al* [24]	TQWT and LS-SVM	Fuzzy entropy	84.67
Sharma *et al* [25]	EMD and LS-SVM	Entropy	87
Das *et al* [26]	EMD, DWT and KNN	Log energy entropy	89.4
Sharma *et al* [27]	TQWT and LS-SVM		94.25
Gupta *et al* [28]	FAWT and LS-SVM	Multiple entropies	94.41
Proposed method	OAS and ELM		98.94

In figure 2.7 the TB kernel started with very low accuracy. We observed a consistent trend up to 200 nodes and accuracy then remained constant for 200 and 210 nodes. Similarly, RBF accuracy increased to 120, remained stable for 120 and 130 nodes, and eventually showed an increase at 200 nodes and remained constant for 200 and 210 nodes. For SINE, accuracy showed a persistent increase from 110 to 200 without fluctuations and remained constant for 200 and 210 nodes. However, SIG showed a continuous increase up to 120, remained stable for 120 and 130 nodes, and eventually showed an increase in accuracy at 200 nodes and remained constant for 200 and 210 nodes.

Table 2.4 shows the performance comparison analysis of existing works against the proposed model. Sharma *et al* investigated an EMD-based entropy feature with LS-SVM as the classifier variant and achieved an accuracy of 84% and 87%, and using TQWT and LS-SVM classifier achieved an accuracy of 94.25% [22, 25, 27]. Bhattacharyya *et al* used TQWT with the LS-SVM classifier and achieved an accuracy of 84.67% [24]. Similarly, Das *et al* used an EMD and DWT extracted LEE feature, fed to the LS-SVM classifier, and obtained an accuracy of 89.4% [26]. Gupta *et al* used FAWT features fed to LS-SVM classifier and achieved an accuracy of 94.41% for classifying F and NF EEG signals [28]. The proposed model uses OAS with an ELM classifier and achieves an accuracy of 98.94%.

2.4 Conclusion

The spontaneously varying nature of EEG signals makes their analysis complicated. Therefore, we have developed an effective hybrid focal and non-focal detection model to extract hidden information and accurately identify focal EEG signals. The time–frequency distribution provided by the Wigner–Ville distribution effectively detects rapid variations in the EEG signals. Due to this, features of the seventh fully connected layer of the pre-trained AlexNet model have effectively extracted the discriminable features. The classification report of the extreme learning machine classifier shows that our developed model yielded the highest performance of accuracy, sensitivity, specificity, F1 score, Matthews correlation coefficient, and precision of 98.94%, 98.72%, 99.17%, 98.94%, 97.89%, and 99.17%, respectively. The performance report shows that our developed model is effective, accurate, and robust as it obtained the highest performance and developed using a ten-fold cross-validation technique. Our developed model has some limitations as it is tested on a single EEG dataset and has not been tested using the leave-one-subject-out validation technique. In the future, we will present a more robust and comprehensive model, including model explainability and uncertainty quantification for other neurological disorders.

References

[1] Khare S K, Gaikwad N B and Bajaj V 2022 VHERS: a novel variational mode decomposition and Hilbert transform-based EEG rhythm separation for automatic ADHD detection *IEEE Trans. Instrum. Meas.* **71** 1–10

[2] Khare S K and Bajaj V 2021 A CACDSS for automatic detection of Parkinson's disease using EEG signals *2021 Int. Conf. on Control, Automation, Power and Signal Processing (CAPS)* pp 1–5

[3] Acharya U R, Sree S V, Swapna G, Martis R J and Suri J S 2013 Automated EEG analysis of epilepsy: a review *Knowl Based Syst.* **45** 147–65

[4] Khare S K and Bajaj V 2021 A self-learned decomposition and classification model for schizophrenia diagnosis *Comput. Methods Programs Biomed.* **211** 106450

[5] Arunkumar N *et al* 2017 Classification of focal and non focal EEG using entropies *Pattern Recognit. Lett.* **94** 112–7

[6] Andrzejak R G, Schindler K and Rummel C 2012 Nonrandomness, nonlinear dependence, and nonstationarity of electroencephalographic recordings from epilepsy patients *Phys. Rev.* E **86** 46206

[7] Taran S, Bajaj V, Sharma D, Siuly S and Sengur A 2018 Features based on analytic IMF for classifying motor imagery EEG signals in BCI applications *240 Measurement* **116** 68–76

[8] Taran S and Bajaj V 2018 Clustering variational mode decomposition for identification of focal EEG signals *IEEE Sens. Lett.* **2** 1–4

[9] Taran S, Bajaj V and Siuly S 2017 An optimum allocation sampling based feature extraction scheme for distinguishing seizure and seizure-free EEG signals *Health Inf. Sci. Syst.* **5** 1–7

[10] Ullo S L, Khare S K, Bajaj V and Sinha G R 2020 Hybrid computerized method for environmental sound classification *IEEE Access* **8** 124055–65

[11] Siuly S and Li Y 2015 Discriminating the brain activities for brain–computer interface applications through the optimal allocation-based approach *Neural Comput. Appl.* **26** 799–811

[12] Pareta A, Taran S, Bajaj V and Sengur A 2019 Automatic environment sounds classification using optimum allocation sampling *4th Int. Conf. on Robotics and Automation Engineering (ICRAE)* pp 69–73

[13] Sultan Qurraie S and Ghorbani Afkhami R 2017 ECG arrhythmia classification using time frequency distribution techniques *Biomed. Eng. Lett.* **7** 325–32

[14] Abeysekera R and Boashash B 1989 Time-frequency domain features of ECG signals: their application in P wave detection using the cross Wigner–Ville distribution *Int. Conf. on Acoustics, Speech, and Signal Processing* pp 1524–7

[15] Sharma R R, Meena P and Pachori R B 2020 Enhanced Time-Frequency Representation Based on Variational Mode Decomposition and Wigner-Ville Distribution *Recent Trends in Image and Signal Processing in Computer Vision* vol 1124. (Berlin: Springer) https://doi.org/10.1007/978-981-15-2740-1_18

[16] Pachori R B and Nishad A 2016 Cross-terms reduction in the Wigner–Ville distribution using tunable-Q wavelet transform *Signal Process.* **120** 288–304

[17] Dhok S, Pimpalkhute V, Chandurkar A, Bhurane A A, Sharma M and Acharya U R 2020 Automated phase classification in cyclic alternating patterns in sleep stages using Wigner–Ville distribution based features *Comput. Biol. Med.* **119** 103691

[18] LeCun Y, Bottou L, Bengio Y and Haffner P 1998 Gradient-based learning applied to document recognition *Proc. IEEE* **86** 2278–324 pp

[19] Alom M Z *et al* 2018 The history began from AlexNet: a comprehensive survey on deep learning approaches arXiv:1803.01164

[20] Khare S K, Bajaj V and Acharya U R 2021 SPWVD-CNN for automated detection of schizophrenia patients using EEG signals *IEEE Trans. Instrum. Meas.* **70** 1–9

[21] Khare S K, Bajaj V, Taran S and Sinha G R 2022 Multiclass sleep stage classification using artificial intelligence based time-frequency distribution and CNN *Artificial Intelligence-Based Brain-Computer Interface* (Amsterdam: Elsevier) pp 1–21

[22] Sharma R, Pachori R B and Acharya U R 2015 An integrated index for the identification of focal electroencephalogram signals using discrete wavelet transform and entropy measures *Entropy* **17** 5218–40

[23] Krizhevsky A, Sutskever I and Hinton G E 2017 ImageNet classification with deep convolutional neural networks *Commun. ACM* **60** 84–90

[24] Bhattacharyya A, Pachori R B and Acharya U R 2017 Tunable-q wavelet transform based multivariate sub-band fuzzy entropy with application to focal EEG signal analysis *Entropy* **19** 1–14

[25] Sharma R, Pachori R B and Gautam S 2014 Empirical mode decomposition based classification of focal and non-focal seizure EEG signals *Proc. Int. Conf. Medical Biometrics* pp 135–40

[26] Das A B and Bhuiyan M I H 2016 Discrimination and classification of focal and non-focal EEG signals using entropy-based features in the EMD-DWT domain *Biomed. Signal Process. Control* **29** 11–21

[27] Sharma R, Kumar M, Pachori R B and Acharya U R 2017 Decision support system for focal EEG signals using tunable-q wavelet transform *J. Comput. Sci.* **20** 52–60

[28] Gupta V, Priya T, Yadav A K, Pachori R B and Acharya U R 2017 Automated detection of focal EEG signals using features extracted from flexible analytic wavelet transform *Pattern Recognit. Lett.* **94** 180–8

[29] Demir F, Bajaj V, Ince M C, Taran S and Şengür A 2019 Surface EMG signals and deep transfer learning-based physical action classification *Neural Comput. Appl.* **31** 8455–62

[30] Bajaj V, Taran S, Tanyildizi E and Sengur A 2019 Robust approach based on convolutional neural networks for identification of focal EEG signals *IEEE Sens. Lett.* **3** 1–4

[31] Bajaj V, Taran S, Khare S K and Sengur A 2020 Feature extraction method for classification of alertness and drowsiness states EEG signals *Appl. Acoust.* **163** 107224

[32] Krishna A H, Sri A B, Priyanka K Y V S, Taran S and Bajaj V 2019 Emotion classification using EEG signals based on tunable-Q wavelet transform *IET Sci., Meas. Technol.* **13** 375–80

[33] Taran S and Bajaj V 2018 Drowsiness detection using adaptive hermite decomposition and extreme learning machine for electroencephalogram signals *IEEE Sens. J.* **18** 8855–62

[34] Ahmad S, Agrawal S, Joshi S, Taran S, Bajaj V, Demir F and Sengur A 2020 Environmental sound classification using optimum allocation sampling based empirical mode decomposition *Physica* A *537 122613*

[35] Taran S and Bajaj V 2018 Rhythm-based identification of alcohol EEG signals *IET Sci., Meas. Technol.* **12** 343–9

[36] Huang G-B, Zhu Q-Y and Siew C-K 2006 Extreme learning machine: theory and applications *Neurocomputing* **70** 489–501

[37] Tan P, Sa W and Yu L 2016 Applying extreme learning machine to classification of EEG BCI *IEEE Int. Conf. on Cyber Technology in Automation, Control, and Intelligent Systems (CYBER)* pp 228–32

[38] Kumar S T S and Kasthuri N 2019 EEG seizure classification based on exploiting phase space reconstruction and extreme learning *Cluster Comput.* **22** 11477–87

[39] Khare S K, Nishad A, Upadhyay A and Bajaj V 2020 Classification of emotions from EEG signals using time-order representation based on the S-transform and convolutional neural network *Electron. Lett.* **56** 1359–61

[40] Khare S K and Bajaj V 2020 Time–frequency representation and convolutional neural network-based emotion recognition *IEEE Trans Neural Netw. Learn. Syst.* **32** 2901–9

[41] Khare S K, Gaikwad N and Bokde N D 2022 An intelligent motor imagery detection system using electroencephalography with adaptive wavelets *Sensors* **22** 8128

[42] Khare S K, Bajaj V and Sinha G R 2020 Automatic drowsiness detection based on variational non-linear chirp mode decomposition using electroencephalogram signals *Modelling and Analysis of Active Biopotential Signals in Healthcare* (Bristol: IOP Publishing) vol 1

IOP Publishing

Data Analytics for Intelligent Systems
Techniques and solutions
Sachin Taran, Chhavi Dhiman and Manjeet Kumar

Chapter 3

Multi-channel EEG-based affective emotion identification using a dual-stage filtering approach

Kranti S Kamble and Joydeep Sengupta

Machine learning (ML)-based classifiers have produced encouraging results for electroencephalogram (EEG)-based affective emotion detection. To identify emotions, a noise-free desirable frequency band of the EEG data is required to be retrieved. In the proposed work, a dual-stage correlation and instantaneous frequency (CIF) threshold method is proposed for retrieval of the noise-free desirable EEG frequency band for affective emotion detection. First, the noisy intrinsic mode functions (IMFs) of the raw EEG signals generated from the empirical mode decomposition technique are eliminated by applying the correlation thresholding criterion. Second, using non-linear chirp variational mode decomposition to decompose first-stage noise-free EEG signals and applying an IF-based filtering criterion to the modes, the desired frequency bands (4–30 Hz) are extracted. To divide emotions into arousal, valence, and dominance dimensions, the collected features from the filtered three frequency bands are fed to ML-based classifiers. This work also demonstrates the superiority of ensemble machine learning (EML) classifiers over conventional machine learning (CML) classifiers. The best F1-scores of 84.53%, 76.24%, and 89.16% were reported by random forest with differential entropy features for arousal, valence, and dominance, respectively. Similarly, the respective average F1-score of the three EML classifiers is ~2.30%, ~7.49%, and ~2.65% higher compared to the three CML classifiers. In conclusion, the suggested CIF-based filtering strategy helps to identify affective emotions within the context of EML classifiers.

3.1 Introduction

Human emotions are essential for everyday communication and decision-making. To create better human–machine interactions, reliably identifying emotions from

EEG data has become essential. Human emotions can be expressed and identified accurately by EEG signals [1], because they offer accurate information on several subjective aspects of mental states [2]. Generally, emotions are categorized into discrete (joy, fear, sadness, and disgust) [3] and multi-dimension emotions (arousal, valence, and dominance) [4]. Arousal shows how emotions can be active or passive, valence describes how strongly a sensation is pleasant or unpleasant, and dominance affects whether a person feels in control or powerless. This study used these three dimensions to categorize human emotions.

To categorize emotions from EEG data, researchers have so far demonstrated various machine learning and deep learning techniques [5–9]. To classify emotions into arousal and valence dimensions, power spectral density (PSD) of different frequency bands was extracted and fed into a support vector machine (SVM) [5]. The Fourier–Bessel series expansion-based empirical wavelet transform (FBSE-EWT) approach with SVM architecture was suggested by Pachori *et al* [7] for cross-subject affective emotion detection on the DEAP database. However, their expanded study of FBSE-EWT using ensemble bagged trees surpassed the initial approach [6]. Several features were obtained by the study [8] utilizing the flexible analytic wavelet transform (FAWT) and then used with a k-nearest neighbor (kNN) classifier for emotion identifications. Several papers have reported emotion detection like this utilizing ML classifiers [10–12].

The majority of this research has used traditional ML-based classifiers such as KNN or SVM, which consistently produced low emotion detection results. Therefore, to improve overall emotion identification accuracy, ensemble ML (EML)-based classifiers are used [7, 13, 14]. The RF classifier achieves valence and arousal accuracies of 79.99% and 79.95% by decomposing the EEG signal using the FAWT technique [7]. Sangnark *et al* [13] also used the RF classifier to retrieve music preferences from EEG signals, although for Kamble *et al* [14] the method was more effective. In a different work, Subasi *et al* [15] extracted features using a tunable-Q-wavelet transform and applied the rotating forest ensemble method to the SEED dataset to achieve an overall emotion detection accuracy of 93%.

To deal with complex signals, various non-stationary signal processing methods are also suggested, such as the wavelet transform, multiwavelet transform, and time–frequency distributions such as empirical mode decomposition (EMD), variational mode decomposition (VMD), and non-linear chirp variational mode decomposition (NCVMD). Additionally, it has been documented in the literature that adaptive EMD and VMD methods are used extensively to analyze non-linear and non-stationary signals. EMD is an approach that places no restrictions on the linearity and stationarity of signal. VMD out-performs the well-liked wavelet-based decompositions and EMD for the segregation of multi-component signals [16]. The VMD's main issue is that it still uses a frequency-domain formulation and is unable to deconstruct wide-band modes with overlapping spectra. A further limitation of the VMD is that it can only split the modes and is unable to extract mode properties directly such as instantaneous amplitude (IA) and instantaneous frequency (IF) [17]. NCVMD is therefore created to segment the EEG signals into narrow-band modes that can be easily distinguishable in frequency/time–frequency regions [18].

Applying demodulation techniques, wide-band non-linear chirp EEG signals can be transformed into narrow-band EEG signals.

To deal with non-stationary EEG signals, this study integrates the benefits of EMD and NCVMD. The stationary properties are improved using EMD [16], while NCVMD uses that signal to create the appropriate frequency oscillatory EEG signal [18]. Motivated by this, we aim to develop an ML-based affective emotion detection system using a dual-stage correlation–instantaneous frequency-based (CIF) filtering approach within the context of EML classifiers. The overall affective emotion identification system is built up through dual-stage CIF filtering, feature extraction, and an emotion classification process. We claim that EML systems are superior at predicting affective emotions than CML classifiers.

3.2 Method

3.2.1 Database

A multimodal DREAMER database with 14-channel EEG signals was used in this study [19]. The database was collected from 23 (14 male and 9 female) healthy subjects with ages ranging from 22 to 33 years old. To elicit the participants' emotions, 18 movie clips were employed as audio-visual stimuli. The nine emotions that were the focus of the movie clips were—amusement, anger, disgust, excitement, fear, happiness, calmness, sadness, and surprise. All movie clips lasted 65–393 s, which is thought to be long enough to evoke just one emotion. The participants evaluated each movie clip on a 1–5 scale for arousal, valence, and dominance. High arousal, valence, or dominance was defined as a sample score greater than 3 and vice versa. All EEG signals were captured at a 128 Hz sampling frequency. A thorough explanation of DREAMER is provided in [19].

3.2.2 Dual-stage CIF filtering with EMD-NCVMD integration

To achieve the appropriate frequency component of the input raw EEG data and to remove noise, this study used EMD and NCVMD separately for CIF filtering, termed 'dual-stage CIF filtering', as shown in figure 3.1. A similar strategy of employing CIF filtering with two distinct approaches consecutively for noise-free desired frequency band EEG signals has been attempted in a previous study [16]. First filtering allows removing the noisy intrinsic mode functions (IMFs) by applying correlation thresholding on the IMFs generated from raw EEG signals. The remaining IMFs are used to generate the noise-free EEG signal, as illustrated in figure 3.1. Second filtering allows obtaining denoised desired frequency bands through the modes obtained by NCVMD by setting IF-based filtering criteria. Additionally, the IF-based filtering norm filters out unwanted frequency components from NCVMD modes. The following section contains information on two-stage filtering.

3.2.2.1 EMD-noise filtering

EMD is applied on all channels of raw EEG signals independently and divides the signals into numerous amplitude–frequency modulated (AM–FM) oscillatory

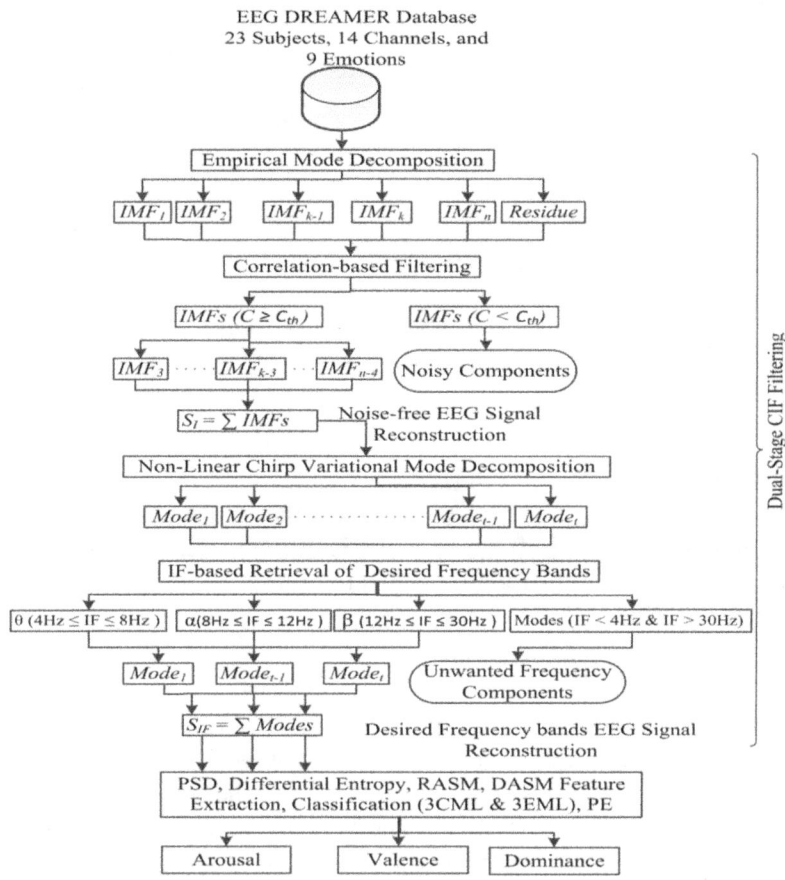

Figure 3.1. Dual-stage filtering approach for affective emotion identification.

patterns known as IMFs. These IMFs should satisfy two requirements: (i) there must be an equal or smaller difference between the frequencies of zero crossings and the extrema, and (ii) local maxima and minima must always determine the mean envelope value of zero. IMFs can express a raw EEG signal $h(t)$ as

$$h(t) = r_k + \sum_{i=1}^{J} g_i(t), \qquad (3.1)$$

where r_k, J, and g_i stand for the residue, number of retrieved IMFs, and ith IMF, respectively. IMFs represent the informative or noisy portion of the original EEG signal. Noisy parts have weaker correlations and matching to the original EEG signals than informative parts [20]. Therefore, to eliminate such components, a correlation criterion with the best correlation threshold C_{th} is applied to IMFs:

$$C_{th} = \frac{\max(C_j)}{10 \times \max(C_j) - 3} \quad j = 1, 2..., n, \qquad (3.2)$$

where C_j denotes the correlation coefficient (CC) between the original EEG signal and ith IMF, and $\max(C_j)$ stands for the maximum CC among J IMF.

3.2.2.2 NCVMD-desired frequency component filtering

The appropriate frequency bands (4–30 Hz) are obtained for affective emotion identification using IF-based filtering of modes derived from NCVMD [19]. NCVMD aims to segment the non-linear chirp signal into distinct signal modes and separate them into the desired frequency bands, which are termed non-linear chirp modes (NCMs). The collected signal is made up of various NCMs which are functions of AM–FM and can be stated as the following:

$$m(t) = \sum_{j=1}^{M} a_j(t) \quad \cos\left(2\pi \int_0^t f_j(s)\, ds + \phi_j\right) \quad , \tag{3.3}$$

where M denoted modes ($M = 5$ in this case), $a(t) > 0$, $f(t) > 0$ stands for the IA and IF of NCM, respectively, and ϕ represents the initial phase. Compute such IF for all modes which lie in frequency bands of θ (4–8 Hz), α (8–12 Hz), and β (12–30 Hz) then filter out the remaining IFs to obtain the desired frequency band denoised EEG signals.

3.2.3 Feature extraction

The features of the final 60 s of each video clip are retrieved and analyzed as this is adequate to induce the desired emotions [19]. It is well known that a human's emotional state and the PSDs of EEG signals from the desired frequency bands (4–30 Hz) are correlated [21]. Positive emotions are better represented in higher-frequency EEG signals than negative ones, according to Soleymani *et al* [22] (high and low valence, respectively). Significant correlations among valence and EEG signals were found across entire frequency ranges by Koelstra *et al* [23]. Additionally, they discovered positive correlations between arousal and elevated beta power. However, some researchers found that the alpha band power and overall arousal level are inversely related [24]. The PSD of each denoised EEG signal mode is estimated by Welch's overlapped segment average estimator using a 256-sample window with a 128-sample overlap. Using the denoised 14-channel EEG signals, logarithmic PSDs are extracted, leading to a total of 210 (3 bands × 5 modes × 14 channels) features.

Along with PSD, we also retrieved differential entropy (DE), differential asymmetry (DASM), and rational asymmetry (RASM) features. DE can distinguish between low- and high-frequency energy of EEG pattern and is stated as

$$DE = -\int f(z) \log(f(z)) dz \quad . \tag{3.4}$$

For a given EEG sequence, the DE is calculated for the θ, α, and β frequency bands. Hence, the 14-channel band-separated EEG signal has DE feature with 210 (3 bands × 5 modes × 14 channels) dimensions. In the processing of emotions, asymmetrical brain activity (lateralization going left–right) appears to be useful [25]. Since there were 14 pairs of hemispheric asymmetry electrodes (left hemisphere: AF3, FC5, F3, F7, T7, P7, and O1; right hemisphere: AF4, FC6, F4, F8, T8, P8, and O2), the differences and ratios among DE features are used to determine the DASM and RASM features as follows, respectively,

$$DASM = DE\left(Z_{\text{left}}\right) - DE\left(Z_{\text{right}}\right) \qquad (3.5)$$

$$RASM = DE\left(Z_{\text{left}}\right)/ DE\left(Z_{\text{right}}\right), \qquad (3.6)$$

where Z_{left} and Z_{right} stand for left and right hemisphere electrode pairs. Therefore, DASM and RASM have features with 105 dimensions (3 bands $\times$ 5 modes $\times$ 7 channels).

After combining each feature from five modes and three bands, the resultant feature vectors produced are F_{PSD}/DE and $F_{\text{DASM/RASM}}$. Let $F_{\theta im_1}$, $F_{\theta im_2}$, $F_{\theta im_3}$, $F_{\theta im_4}$, and $F_{\theta im_5}$ be the log(PSD)/DE/DASM/RASM feature of the ith channel of θ frequency band of EEG signal, PSD/DE: $i = 1, 2, ..., 14$ or DASM/RASM: $i = 1, 2, ..., 7$ for five modes. The same applies to α and β frequency bands. The final feature vectors produced are $F_{\text{PSD/DE}} = \left[F_{\theta 1m_1} F_{\alpha 1m_1} F_{\beta 1m_1} ... F_{\theta 1m_5} F_{\alpha 1m_5} F_{\beta 1m_5} ... F_{\theta 14m_5} F_{\alpha 14m_5} F_{\beta 14m_5} \right]$ and $F_{\text{DASM/RASM}} = \left[F_{\theta 1m_1} F_{\alpha 1m_1} F_{\beta 1m_1} ... F_{\theta 1m_5} F_{\alpha 1m_5} F_{\beta 1m_5} ... F_{\theta 7m_5} F_{\alpha 7m_5} F_{\beta 7m_5} \right]$.

3.2.4 Classification

For affective emotion identification, the SVM, kNN, logistic regression (LR), bagging, random rotation forest (RRF), and RF ML-based classifiers were used in this study. The entire feature vectors were split into training and test sets applying the stratified ten-fold cross-validation (CV) methodology. Classification samples are divided into ten independent subgroups at random, with the process being carried out ten times, resulting in nine training subsets and one testing subset. The average categorization performance across all ten-folds is estimated in the end. The overall efficacy of ML-based affective emotion detection is assessed using three performance evaluation metrics, including accuracy, F1-score, and AUCs. Using the five-fold CV RandomSearchCV technique, the hyperparameters of ML-based classifiers are tuned. Table 3.1 gives the list of hyperparameters.

3.3 Result and discussion

The supervised binary classification was conducted using dual-stage CIF filtering for EEG-based affect emotion identification. The binary classification was categorized as low/high arousal, valence, and dominance. Ground truth was the participant's self-reported ratings from 1–5 score. The mid-point threshold causes uneven class distribution for some participants for arousal, valence, and dominance dimensions. Therefore, the F1-score was also evaluated to address asymmetric class distributions along with accuracy and AUC.

In this study, the classification was done independently using three CML classifiers—SVM with an RBF kernel, kNN, and LR—and three EML classifiers —the bagging, rotation random forest (RRF), and RF classifiers. The affective

Table 3.1. Complete list of hyperparameters for ML-based classifiers.

Classifiers	Hyperparameters	Grid of hyperparameters	Optimized parameter
	Kernel	['rbf', 'linear', 'poly']	rbf
	Gamma	[0.001, 0.01, 0.1, 1]	1
SVM	C	[0.001, 0.01, 0.1, 1, 2]	2
	Degree	[1, 2, 3]	1
	Penalty	[11, 12]	L2
KNN	p	[1, 2]	2
	n_neighbors	[1, 2, 3, 4, 5, 6, 7, 8, 9]	3
	leaf_size	[1, 2, 3, 4, 5, 6, 7, 8, 9]	8
LR	C	[1e−04, 7.7e−04, 5.9e−03, 4.6e−02, 3.5e−01, 2.8e+0, 2.2e+01, 1.7e+02, 1.3e+03, 1.0e+04]	1.7e+02
Bagging	n_estimators	[100, 200, 500, 1000]	500
RRF	n_estimators	[10, 50, 70, 100, 125,200]	200
	k	[3, 5, 7]	3
	n_estimators	[100, 200, 500, 1000]	1000
	min_weight_fraction_leaf	[0.1, 0.01]	0.1
RF	min_samples_split	[2, 5, 10]	5
	min_samples_leaf	[2, 5, 10]	5
	max_leaf_nodes	[20, 40]	10
	Max_depth	[3, 4, 5]	4

emotion detection system was built utilizing an open-source Python Library called Scikit-learn.

Table 3.2 shows the total average performance metrics across all participants for three dimensions. Notably, In table 3.2 EEG data from three frequency bands are directly concatenated. We compare the performance of four features: PSD, DE, DASM, and RASM. From the results, it can be observed that the DE feature achieved the best classification average among all the features.

For arousal, valence, and dominance, the best average accuracies with DE features over three EML are, respectively, 7.84% (77.65% versus 69.81%), 10.17% (69.92% versus 59.75%), and 3.02% (77.83% versus 74.81%) higher than the average accuracies over the three CML classifiers. Similarly, the respective F1-score is 2.30% (83.78% versus 81.48%), 7.49% (74.65% versus 67.16%), and 2.65% (87.30% versus 84.65%) higher compared to the three CML algorithms.

We can focus more on the F1-score because the data has an unequal distribution of classes. The EML-based RF classifier reports the highest F1-score for the arousal, valence, and dominance dimensions of 84.53%, 76.24%, and 89.16%, respectively, while, the corresponding highest accuracies are 78.79%, 72.64%, and 78.68%, respectively. This corroborates our claim that EML classifiers may predict affective emotion identification more precisely than CML classifiers.

Table 3.2. Comparison of classification performance in terms of (PE_Metrics ± SD) of CML and EML algorithms using (a) PSD, (b) DE, (c) DASM, and (d) RASM. The values reported in bold represent amount of improvement of EML over CML algorithm.

Feature	Type of algorithm	Algorithms	Accuracy (%)			F1-score (%)		
			Arousal	Valence	Dominance	Arousal	Valence	Dominance
(a)								
	CML	SVM	72.41 ± 1.01	69.81 ± 1.58	77.25 ± 1.05	82.83 ± 0.74	74.76±1.29	87.16±0.67
		kNN	63.93 ± 7.66	45.06 ± 7.01	68.51 ± 6.13	76.72 ± 5.43	49.27±9.07	80.44±4.01
		LR	72.16 ± 1.09	70.80 ± 4.61	77.25 ± 1.05	82.77 ± 0.72	74.37±3.1	87.16±0.67
PSD	Average		69.5	61.89	74.33	80.77	66.13	84.92
	EML	Bagging	73.04 ± 4.39	71.16 ± 8.28	76.76 ± 1.8	79.82 ± 2.96	73.83±7.43	86.21±1.27
		RRF	68.99 ± 3.65	59.54 ± 5.18	75.3 ± 2.51	81.0 ± 2.62	70.46±3.98	85.62±1.59
		RF	74.48 ± 2.21	71.99 ± 2.92	78.35 ± 1.81	83.99 ± 0.68	75.94±2.04	88.86±2.99
	Average		73.76	67.56	76.80	81.60	73.41	86.89
Improvement of EML over CML (%)			**4.26**	**5.67**	**2.47**	**0.83**	**7.28**	**1.97**
(b)								
	CML	SVM	72.16 ± 1.09	**61.52 ± 1.38**	74.08 ± 3.83	**83.83 ± 0.74**	**75.87±0.92**	83.26±1.44
		kNN	65.13 ± 7.69	57.69 ± 7.86	73.12 ± 4.92	76.79 ± 6.1	51.30±7.04	83.55±3.27
		LR	72.16 ± 1.09	60.04 ± 3.91	77.25 ± 1.05	83.83 ± 0.74	74.31±2.71	87.16±2.63
DE	Average		69.81	59.75	74.81	81.48	67.16	84.65
	EML	Bagging	78.79 ± 3.43	68.26 ± 7.43	75.17 ± 3.11	84.53 ± 2.44	74.70±5.26	86.98±1.92
		RRF	76.76 ± 3.31	68.87 ± 6.05	77.86 ± 4.21	82.90 ± 2.39	73.03±4.03	85.77±2.63
		RF	77.40 ± 1.01	69.01 ± 3.25	78.68 ± 1.05	83.92 ± 0.62	75.24±2.26	**89.16±2.63**
	Average		77.65	68.71	77.23	83.78	74.32	87.30
Improvement of EML over CML (%)			**7.84**	**9.96**	**2.42**	**2.30**	**7.16**	**2.65**

(c)

DASM	CML	SVM	72.67 ± 4.94	60.29 ± 1.62	76.02 ± 3.33	83.99 ± 0.68	75.15 ± 0.89	86.26 ± 1.94
		kNN	61.71 ± 6.98	53.99 ± 10.09	68.50 ± 8.21	74.14 ± 4.76	58.42 ± 9.62	80.53 ± 5.71
		LR	69.40 ± 1.27	61.26 ± 1.05	72.67 ± 4.94	82.01 ± 0.52	75.90 ± 0.62	83.99 ± 3.38
		Average	67.92	58.51	72.39	80.04	69.82	83.59
	EML	Bagging	74.32 ± 5.18	59.56 ± 5.02	74.08 ± 3.83	84.47 ± 3.91	76.08 ± 4.22	84.91 ± 2.44
		RRF	72.98 ± 3.89	57.88 ± 8.42	76.02 ± 3.33	81.99 ± 2.74	69.31 ± 6.81	87.16 ± 0.67
		RF	71.53 ± 2.43	62.02 ± 1.24	77.25 ± 1.05	81.91 ± 1.44	75.22 ± 1.27	88.03 ± 0.53
		Average	72.94	59.82	75.78	82.79	73.53	86.7
Improvement of EML over CML (%)			**5.02**	**1.30**	**3.38**	**2.75**	**3.70**	**3.11**

(d)

RASM	CML	SVM	72.16 ± 1.09	58.84 ± 2.97	77.00 ± 1.98	81.26 ± 3.02	73.65 ± 2.41	86.97 ± 1.21
		kNN	63.70 ± 6.05	47.18 ± 9.01	72.15 ± 3.82	76.34 ± 4.53	53.98 ± 7.59	82.86 ± 2.19
		LR	72.40 ± 2.23	51.32 ± 7.46	75.07 ± 3.36	83.99 ± 0.68	68.12 ± 3.97	86.30 ± 1.56
		Average	69.42	52.44	74.74	80.53	67.25	85.37
	EML	Bagging	73.01 ± 4.08	54.22 ± 5.94	76.03 ± 2.62	82.34 ± 2.22	74.56 ± 1.60	87.16 ± 0.67
		RRF	69.26 ± 3.68	59.56 ± 2.07	76.55 ± 2.25	81.27 ± 2.95	64.78 ± 6.20	85.63 ± 2.01
		RF	**75.34 ± 2.31**	**60.05 ± 1.91**	**79.56 ± 3.69**	**84.32 ± 1.56**	**75.03 ± 1.51**	**89.84 ± 2.45**
		Average	72.53	57.94	77.37	82.64	71.45	87.54
Improvement of EML over CML (%)			**3.11**	**5.49**	**2.63**	**2.11**	**4.2**	**2.16**

Even though asymmetric features (DASM and RASM) have fewer feature dimensions than PSD and DE features, the outcomes are still comparable. This finding is also consistent with the study in [25].

Figure 3.2 compares the ROC plots for PSD, DE, DASM, and RASM features using the optimal ensemble RF algorithms. The AUCs are evaluated with a 95% confidence interval for all three dimensions. From figures 3.2 (A)–(C), the highest average AUCs among all features given by DE for arousal and dominance dimensions are 0.82 (95% CI = 0.81–0.83), and 0.84 (95% CI = 0.83–0.85), respectively, whereas RASM gives 0.79 (95% CI = 0.79–0.80) for valence dimension. This supports and validates our claim that ensemble algorithms may identify emotions more accurately than conventional algorithms.

The proposed ML-based affective emotion identification system is compared with studies that used DREAMER and other databases, as shown in table 3.3. In all the studies the identical features of the PSD are applied to the SVM classifier to identify the affective emotions. The proposed study performs better than the state-of-the-art with F1-scores of 82.83%, 74.76%, and 87.16% for arousal, valence, and dominance,

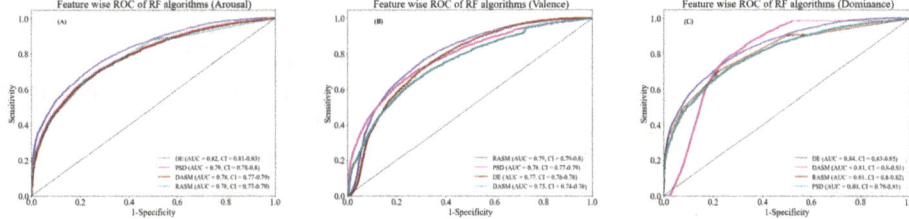

Figure 3.2. ROC curves comparing extracted features of optimal ensemble RF algorithms for affective emotion identification for the (A) arousal, (B) valence, and (C) dominance dimensions.

Table 3.3. Comparison table for accuracy and F1-score for affect emotion identification for DREAMER and other databases which applied PSD for feature extraction analysis. The values reported in bold represent best observed performance of the proposed work.

Modality	Accuracy (%)			F1-score (%)		
	Arousal	Valence	Dominance	Arousal	Valence	Dominance
DREAMER-EEG [19]	62.17	62.49	61.84	57.67	51.84	61.66
DREAMER-fusion [19]	62.32	61.84	61.84	57.50	52.13	61.71
DEAP-EEG [23]	62.00	57.60	—	58.30	56.30	—
DEAP peripheral [23]	57.00	62.70	—	53.30	60.80	—
MAHNOB-HCI [26]	52.40	57.00	—	42.00	56.00	—
Open-BCI [5]	67.44	66.67	—	67.40	64.40	—
E4 [5]	62.28	63.83	—	59.12	62.00	—
Open-BCI & E4 [5]	71.57	65.38	—	69.16	64.28	—
Proposed mode: SVM	**72.41**	**69.81**	**77.25**	**82.83**	**74.76**	**87.16**
Proposed mode: RF	**74.48**	**71.99**	**78.35**	**83.99**	**75.94**	**88.86**

respectively, whereas the ensemble-based RF classifier reports respective F1-scores of 83.99%, 75.94%, and 88.86%. The same has been seen in terms of accuracy in all three dimensions. The results of the study are promising because they surpass past ML-based studies in both accuracy and F1-score metrics. The results support the fact that the two-stage EMD-NCVMD-based CIF filtering approach produces positive outcomes for affect identification systems.

3.4 Conclusion

The efficacy of three ensemble ML classifiers and three conventional ML classifiers were compared to identify the emotional affects from EEG signals in this work. A dual-stage CIF threshold approach that combines EMD and NCVMD techniques together is adopted to obtain the necessary frequency band of noise-free EEG signals. These signals are separated into three frequency bands θ, α, and β. The four different features such as PSD, DE, DASM, and RASM obtained from these bands from five modes are given to ML-based classifiers to categorize emotions into the arousal, valence, and dominance dimensions. The current study's findings demonstrate the advantage of ensemble-based ML classifiers for EEG-driven affective emotion detection by showing that they recognize emotions more precisely than conventional ML classifiers. The ensemble RF technique works better than standard ML-based classifiers, as evidenced by the emotion identification F1-score with DE features of 84.53%, 76.24%, and 89.16% for the arousal, valence, and dominance dimensions, respectively. This supports our claim that ensemble learning-based classifiers are better than traditional ML-based classifiers at identifying affective emotions.

References

[1] Yang Y, Wu Q J, Zheng W-L and Lu B-L 2017 EEG-based emotion recognition using hierarchical network with subnetwork nodes *IEEE Trans. Cogn. Develop. Syst.* **10** 408–19
[2] Li Z, Zhao S, Duan J, Su C-Y, Yang C and Zhao X 2016 Human cooperative wheelchair with brain–machine interaction based on shared control strategy *IEEE/ASME Trans. Mechatron.* **22** 185–95
[3] Ekman P 1992 An argument for basic emotions *Cognit. Emot.* **6** 169–200
[4] Mehrabian A 1996 Pleasure–arousal–dominance: a general framework for describing and measuring individual differences in temperament *Curr. Psychol.* **14** 261–92
[5] Lakhan P *et al* 2019 Consumer grade brain sensing for emotion recognition *IEEE Sens. J.* **19** 9896–907
[6] Anuragi A, Sisodia D S and Pachori R B 2022 EEG-based cross-subject emotion recognition using Fourier–Bessel series expansion based empirical wavelet transform and NCA feature selection method *Inf. Sci.* **610** 508–24
[7] Gupta V, Chopda M D and Pachori R B 2018 Cross-subject emotion recognition using flexible analytic wavelet transform from EEG signals *IEEE Sens. J.* **19** 2266–74
[8] Bajaj V, Taran S and Sengur A 2018 Emotion classification using flexible analytic wavelet transform for electroencephalogram signals *Health Inform. Sci. Syst.* **6** 1–7
[9] Rahman M M, Sarkar A K, Hossain M A and Moni M A 2022 EEG-based emotion analysis using non-linear features and ensemble learning approaches *Expert Syst. Appl.* **207** 118025

[10] Khare S K and Bajaj V 2020 An evolutionary optimized variational mode decomposition for emotion recognition *IEEE Sens. J.* **21** 2035–42

[11] Tuncer T, Dogan S and Subasi A 2021 A new fractal pattern feature generation function based emotion recognition method using EEG *Chaos, Solitons Fractals* **144** 110671

[12] Sharma L D and Bhattacharyya A 2021 A computerized approach for automatic human emotion recognition using sliding mode singular spectrum analysis *IEEE Sens. J.* **21** 26931–40

[13] Sangnark S *et al* 2021 Revealing preference in popular music through familiarity and brain response *IEEE Sens. J.* **21** 14931–40

[14] Kamble K S and Sengupta J 2022 Ensemble machine learning-based affective computing for emotion recognition using dual-decomposed EEG signals *IEEE Sens. J.* **22** 2496–507

[15] Subasi A, Tuncer T, Dogan S, Tanko D and Sakoglu U 2021 EEG-based emotion recognition using tunable Q wavelet transform and rotation forest ensemble classifier *Biomed. Signal Process. Control* **68** 102648

[16] Taran S and Bajaj V 2019 Emotion recognition from single-channel EEG signals using a two-stage correlation and instantaneous frequency-based filtering method *Comput. Methods Programs Biomed.* **173** 157–65

[17] Kamble A, Ghare P and Kumar V 2022 Machine-learning-enabled adaptive signal decomposition for a brain-computer interface using EEG *Biomed. Signal Process. Control* **74** 103526

[18] Chen S, Dong X, Peng Z, Zhang W and Meng G 2017 Nonlinear chirp mode decomposition: a variational method *IEEE Trans. Signal Process.* **65** 6024–37

[19] Katsigiannis S and Ramzan N 2017 DREAMER: a database for emotion recognition through EEG and ECG signals from wireless low-cost off-the-shelf devices *IEEE J. Biomed. Health Inform.* **22** 98–107

[20] Ayenu-Prah A and Attoh-Okine N 2010 A criterion for selecting relevant intrinsic mode functions in empirical mode decomposition *Adv. Adapt. Data Anal.* **2** 1–24

[21] Davidson R J 2003 Affective neuroscience and psychophysiology: toward a synthesis *Psychophysiology* **40** 655–65

[22] Soleymani M, Asghari-Esfeden S, Fu Y and Pantic M 2015 Analysis of EEG signals and facial expressions for continuous emotion detection *IEEE Trans. Affect. Comput.* **7** 17–28

[23] Koelstra S *et al* 2011 DEAP: a database for emotion analysis; using physiological signals *IEEE Trans. Affect. Comput.* **3** 18–31

[24] Barry R J, Clarke A R, Johnstone S J and Brown C R 2009 EEG differences in children between eyes-closed and eyes-open resting conditions *Clin. Neurophysiol.* **120** 1806–11

[25] Zheng W-L and Lu B-L 2015 Investigating critical frequency bands and channels for EEG-based emotion recognition with deep neural networks *IEEE Trans. Auton. Ment. Dev.* **7** 162–75

[26] Soleymani M, Lichtenauer J, Pun T and Pantic M 2011 A multimodal database for affect recognition and implicit tagging *IEEE Trans. Affect. Comput.* **3** 42–55

IOP Publishing

Data Analytics for Intelligent Systems
Techniques and solutions
Sachin Taran, Chhavi Dhiman and Manjeet Kumar

Chapter 4

Variational mode decomposition based entropy features for classification of myopathy, neuropathy, and normal EMG signals

Sukumar Nagineni, Sachin Taran and Kemal Polat

Motor neuron, muscle, and nerve disorder investigations are analysed using electromyogram (EMG) signals. These disorders can be diagnosed with the help of classifying as myopathy, neuropathy, and normal EMG signals. Variational mode decomposition (VMD) is used as the basis of the approach to recognize neuromuscular disorders very efficiently. A signal is decomposed into the band-limited function of modes using an adaptive VMD. Effective entropy-based features are extracted from VMD modes: normalized Shannon entropy, log energy entropy, norm entropy, and sure entropy. All obtained features are simultaneously fed as inputs to the random forest (RF) and distance-weighted k-nearest neighbour (WKNN) classifiers to evaluate the classification performance for the neuromuscular disorder signals. The performance evaluation of the proposed VMD and entropy-based results show the effectiveness of the proposed methodology for the automatic classification of myopathy, neuropathy, and normal EMG signals compared to existing techniques. The proposed methodology based on VMD and entropy has undergone proper performance analysis. It proved more effective in automatically classifying myopathy, neuropathy, and normal EMG signals compared to the existing techniques.

4.1 Introduction

Electromyography is an electrodiagnostic method of recording and evaluating the electrical activities of muscles during contractions and relaxation. This obtained electrical signal activity is called an electromyogram (EMG) [1, 2]. Neuropathy is a nerve disorder that progressively damages nervous control, mainly the peripheral nervous system is affected rather than the central nervous system of the brain and spine. It has two types, mononeuropathy and polyneuropathy. The most common peripheral neuropathy is diabetes, caused by a change in sugar levels resulting in damage to nerve fibres in the feet and legs. It affects about 20 million people in the US. Symptoms are based on the type of nerves damaged and include tingling, hypersensitivity, dizziness, excessive sweating, and diarrhoea. Myopathy is a muscle disorder that affects muscle tissues, resulting in muscle weakness. Myopathy can result from either an acquired cause or an inherited cause. Acquired myopathies, for example caused by alcoholism, have symptoms of stiffness, cramp, and spasm, and inherited myopathies cause muscular dystrophy and inflammation [4]. Nerve conduction studies are useful for a physician in diagnosing muscle function [3, 5–8]. Qualitative analysis of MUAP features in the time domain includes duration, amplitude, area, phases, and turns, and a binary support vector machine (SVM) classifier can help classify myopathy (MYO), neuropathy (NEURO), and normal (NOR) EMG signals [9]. Approximate entropy (ApEn), fractal and correlation dimensions (FD and CD), the Hurst exponent (H), and the largest Lyapunov exponent (LLE) features have been given to classifiers such as the decision tree, fuzzy, and KNN classifiers for classification of MYO, NEURO, and the NOR EMG signals [10].

Singular value decomposition (SVD) and the mean frequency of intrinsic mode function (IMF) features are considered for classifying MYO, NEURO, and the NOR EMG signals using the LS-SVM classifier in [11]. Discrete wavelet transform (DWT) based features, the artificial neural network (ANN) and SVM, and probabilistic neural network classifiers are used to classify MYO, NEURO, and NOR EMG signals in [13]. The autoregressive (AR) method-based features given to decision tree classifiers are CART, random forest, C4.5, and random tree for the classification of MYO, NEURO, and NOR EMG signals in [12]. The continuous wavelet transform (CWT) and multiscale entropy features help automatically classify the abnormal and NOR EMG signals in [14]. AR and DWT methods and an adaptive neuro-fuzzy inference system classifier are used to classify abnormal such as MYO, NEURO and normal EMG signals in [15]. An AR spectrum is given to a wavelet neural network to classify the MYO, NEURO, and NOR EMG signals in [16]. A fast Fourier transform (FFT) based feature set is given to principal components analysis (PCA) to reduce the features' dimensionality, and these PCA coefficients are given to a multilayer perceptron and SVM for classifying the abnormal and the NOR EMG signals in [17]. The vertical visibility algorithm, weight visibility algorithm, and statistical mechanics are used for feature extraction, and these are given to SVM and multilayer perceptron neural networks to classify the MYO, NEURO, and NOR EMG signals in [18, 19]. Extracted MUAP-based features are given to two-stage classifiers, an ANN, and a decision tree, for classifying MYO, NEURO, and NOR EMG signals in [20]. The ensemble empirical mode decomposition method decomposes the signal as amplitude or frequency sub-bands, and time–frequency (T–F) features are extracted using the Hilbert transform. The KNN classifier effectively classified neuromuscular disorder of EMG signals in [21]. Hyperbolic Stockwell transform parameters are optimized with a genetic algorithm for neuromuscular disorder analysis in the T–F plane in [22]. A machine learning pipeline is adopted for distinguishing normal, neuropathy, and myopathy

EMG, which does not rely on the nature of the EMG signal, in [23]. A tunable-Q factor wavelet transform decomposes an input EMG signal into sub-bands. Further, the statistical-based features are extracted and fed to the LS-SVM and KNN for classifying healthy and ALS signals [24]. A weight-vertical visibility method is used to extract features to detect healthy samples, myopathy, and ALS in [25]. Myopathy and neuropathy signal analysis using wavelet coefficients is carried out in [26]. A CWT with KNN and SVM classifiers is used to classify NOR EMG and myopathy signals in [27]. The contribution in this proposed work is as follows:

1. For the first time, a variational mode decomposition based framework is used to classify myopathy, neuropathy, and normal EMG signals.
2. The effective entropy features are extracted from VMD modes and the proposed framework uses a smaller number of features and reduces the complexity of the system.
3. The overall framework gives improved classification results as compared to existing methods.

The methodology is described in section 4.2. The simulation results and discussion related to the classification of NEURO, MYO, and NOR EMG signals are described in section 4.3. The conclusion is presented in section 4.4.

4.2 Methodology

The proposed framework consists of an EMG dataset, a variational mode decomposition based adaptive signal decomposition technique, and entropy features computed from the decomposed modes. These extracted features are fed to random forest (RF) and distance-weighted KNN classifiers for classifying the EMG signals.

4.2.1 Dataset[1]

The recorded dataset of EMG signals is available online [28]. It has three groups: myopathic patients, neuropathic patients, and normal EMG individuals; these data were collected at Harvard Medical School. Signal data acquisition was carried out using a Medelec Synergy N2 EMG monitoring system with the help of a concentric needle electrode of 25 mm inserted into the tibialis anterior muscle of each patient. At the time of recording, the patient was positioned by dorsiflexing the foot softly against resistance. Repositions of a needle electrode until a rapid rise time of unit motor potential were identified. The collection of data lasted several seconds, and after that needle electrode was removed. At that, time patient was in relaxation. The data were collected at a sampling rate of 50 kHz and then down-sampled to 4 kHz and which was digitized by an analogue-to-digital converter. This work uses 111 myopathic, 148 neuropathic, and 51 healthy EMG signals with each class of 1000 samples.

---

[1] https://physionet.org/physiobank/database/emgdb/

4.2.2 Variational mode decomposition

VMD decomposes a real-time function ($f(t)$) into variational modes (u_k). It is an adaptive signal decomposition. Here, k modes (u_k) have occurred concurrently around its centre pulsation (ω_k). The constrained variational problem formulation is helpful in obtaining generalized modes, its optimization problem is expressed as [29, 30]

$$\min_{\{u_k\},\{\omega_k\}} \left\{ \sum_k \left\| \partial_t \left[\left(\delta(t) + \frac{j}{\pi t} \right) \times u_k(t) \right] e^{-j\omega_k t} \right\|_2^2 \right\} \text{ subject to } \sum_k u_k = f. \quad (4.1)$$

The Lagrangian multiplier, Hilbert transform, and the number of variational modes are the more important parameters in the process of decomposition. A Wiener filter is used for updating the modes $u_k(\omega)$ in the VMD using an optimization algorithm known as the alternate direction method of multiplier (ADMM).

The steps involved in a VMD algorithm to decompose an EMG signal are as follows.

Step 1. Predefined the number of modes K, initializing $\{\hat{u}_k^1\}$, $\{\omega_k^1\}$, $\hat{\lambda}^1$, and $n = 0$.

Step 2. Assign $n = n + 1$ as long as $k = 1$: K for $\omega \geqslant 0$; the loop repeats continuously. Update $\hat{u}_k(t)$ in the spectral domain as [29]

$$\hat{u}_k^{n+1}(\omega) \leftarrow \frac{\hat{f}(\omega) - \sum_{i<k} \hat{u}_k^{n+1}(\omega) - \sum_{i>k} \hat{u}_k^n(\omega) + \frac{\lambda^n(\omega)}{2}}{1 + 2\alpha(\omega - \omega_k^n)^2} \quad (4.2)$$

$$\hat{u}_k^{n+1}(t) = R\{\text{ifft}(\hat{u}_k^{n+1}(\omega))\}. \quad (4.3)$$

Update ω_k with

$$\omega_k^{n+1} \leftarrow \frac{\int_0^\infty \omega |\hat{u}_k^{n+1}(\omega)|^2 d\omega}{\int_0^\infty |\hat{u}_k^{n+1}(\omega)|^2 d\omega}. \quad (4.4)$$

The above steps are repeated until k equals K, for n number of total iterations in the loop. The updated Lagrange multiplier (λ) depends on dual ascent for all $\omega \geqslant 0$:

$$\hat{\lambda}^{n+1}(\omega) \leftarrow \hat{\lambda}^n(\omega) + \tau\left(\hat{f}(\omega) - \sum_k \hat{u}_k^{n+1}(\omega)\right). \quad (4.5)$$

Step 3. Repeat the previous step 2 until the following convergence condition is satisfied:

$$\sum_{k=1}^K \frac{\|\hat{u}_k^{n+1} - \hat{u}_k^n\|_2^2}{\|\hat{u}_k^n\|_2^2} < \varepsilon, \quad (4.6)$$

where $\in$, τ, and $\wedge$ represent the tolerance of convergence, time steps of dual ascent, and Fourier transform, respectively. Here, $ifft(.)$, $R(.)$ are the inverse FFT of some signal and the real part of the analytic signal.

4.2.3 Feature extraction

Entropy is a quantitative term for the measurement of disorder in a system. The entropy of a signal x is defined as $E[x]$. The extracted basic four entropies are as described in the following [31–33].

Shannon entropy (E1)

$$E1[x] = -\sum_i x_i^2 \log(x_i^2).$$

(4.7)

Log energy entropy (E2)

$$E2[x] = \sum_i \log(x_i^2), \quad \text{where } \log(0) = 0.$$

(4.8)

Norm entropy (E3)

$$E3[x] = |x|^p.$$

(4.9)

$$\text{Threshold entropy} = \begin{cases} 1, & \text{if } |x_i| > p \\ 0, & \text{elsewhere} \end{cases} \quad p = \sqrt{2 \log_e\left(n \log_2(n)\right)}.$$

(4.10)

Sure entropy (E4)

$$E4[x] = n + \sum_i \min(x_i^2, p^2),$$

(4.11)

where n, p, and x_i are the number of samples, the threshold, and the coefficient of signal x, respectively.

4.2.4 Random forest (RF) classifier

The RF classifier has been proposed by Breiman [34]. It takes an input of a thousand variables, without deletion of input, and then classifies them based on their significance. The ensemble learning algorithms for high-precision applications to obtain better accuracy and robustness so-called one type of ensemble tree classifier. It has an ensemble of classification trees; every individual vote of a tree contributes to the result of classification. In RF, the random vector of each tree is sampled independently and entire trees are distributed equivalently in the forest. Throughout classification, the most popular class and each tree's votes are returned. The RFs consider a few attributes in each split, which is efficient for an extensive database and estimation of internal variable importance [35].

4.2.5 Distance-weighted k-nearest neighbour classifier (WKNN)

In the KNN classifier, the basic idea is to give a query and a set of training samples [36, 37]. A point that is closest to the query is found and then a class label is assigned to that query. Let $T = \{(y_i, z_i)\}_{i=1}^N$ show the training set, y_i the training vector, and z_i corresponding class label. Assumed a query y' and z' an unknown class are assigned in two steps. First, a set of similarly labelled k target neighbours for a query are identified. The set $T' = \{(y_i^{NN}, z_i^{NN})\}_{i=1}^{k}$ is arranged in increasing order in terms of the Euclidean distance $d(y', y_i^{NN})$ between y' and y_i^{NN}:

$$d\left(y', y_i^{NN}\right) = \sqrt{\left(y' - y_i^{NN}\right)^T \left(y' - y_i^{NN}\right)}. \tag{4.12}$$

In the second step, depending on the majority of voting for its nearest neighbours, the query of a class label is predicted by

$$z' = \underset{z}{\mathrm{argmax}} \sum_{\left(y_i^{NN}, z_i^{NN}\right) \in T'} \delta\left(y = y_i^{NN}\right). \tag{4.13}$$

A weighted voting method of KNN is introduced [38] as WKNN. In this, the distance-weighted function decides that closer neighbours have more weight than those far away [39]. The weight w_i of ith nearest query y' is assigned as
The final analysis is based on majority-weighted voting of the classification result of the query is

$$z' = \underset{z}{\mathrm{argmax}} \sum_{\left(y_i^{NN}, z_i^{NN}\right) \in T'} w'_i \times \delta\left(y = y_i^{NN}\right). \tag{4.14}$$

Above, a smaller distance to a neighbour is assigned as more weighted than a larger distance. The furthest neighbour and nearest neighbour are assigned with weights of 0 and 1, respectively. The other neighbours are assigned a distance scaled linearly in between them.

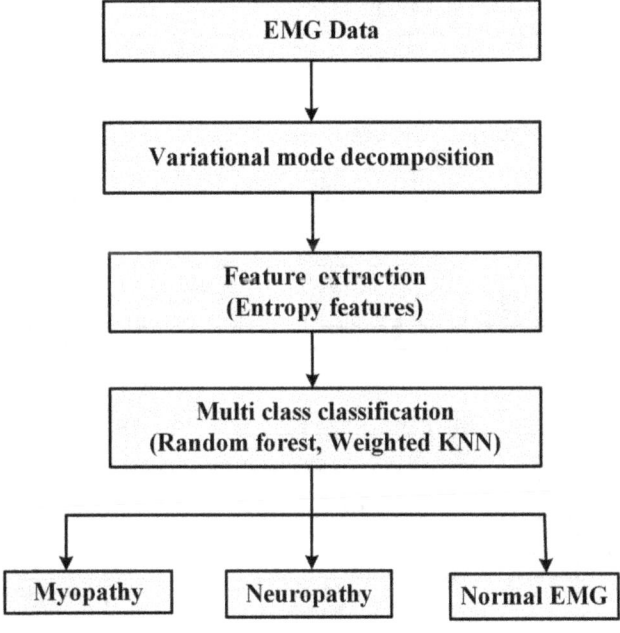

Figure 4.1. VMD-based framework represents classification of MYO, NEURO, and NOR EMG signals.

4.3 Results and discussion

The VMD decomposes an EMG signal into various modes, from which the entropy-based features extracted are the normalized Shannon ($E1$), log energy ($E2$), norm ($E3$), and sure ($E4$) entropy features. It serves as a tool for classifying myopathy, neuropathy, and NOR EMG signals. Figures 4.2(a)–(c) show the decomposed modes of MYO, NEURO, and the NOR EMG signals, respectively. It is observed that the frequency increases as the mode number increases. Entropy feature variations are shown in tables 4.1 and 4.2 with the mean and the standard deviation.

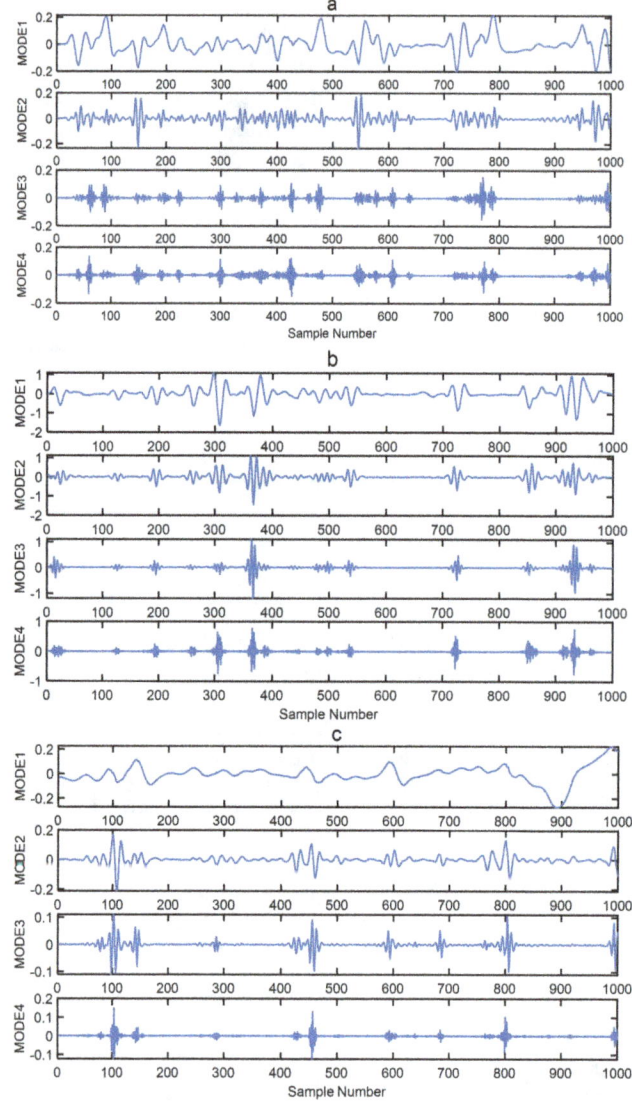

Figure 4.2. Extracted modes from VMD: (a) myopathy, (b) neuropathy, and (c) normal EMG signal.

Table 4.1. Range (mean ± std) values of the Shannon entropy and norm entropy features for the VMD modes.

Mode no.	Signals	Shannon entropy Mean ± std	Norm entropy Mean ± std
Mode 1	Myopathy	−16.595 ± 2.619	4.0218 ± 0.907
	Neuropathy	−39.952 ± 154.032	65.938 ± 137.72
	Normal	−17.0045 ± 11.248	3.8385 ± 3.2065
Mode 2	Myopathy	−8.461 ± 1.304	1.7978 ± 0.3259
	Neuropathy	−32.810 ± 25.796	26.3586 ± 29.75
	Normal	−5.318 ± 3.3978	1.139 ± 1.2009
Mode 3	Myopathy	−3.826 ± 0.668	0.7064 ± 0.1367
	Neuropathy	−13.7365 ± 6.9295	12.762 ± 12.0096
	Normal	−2.125 ± 1.073	0.3870 ± 0.246
Mode 4	Myopathy	−3.190 ± 0.514	0.5989 ± 0.1146
	Neuropathy	−11.2241 ± 6.9373	10.297 ± 10.367
	Normal	−0.8424 ± 0.224	0.143 ± 0.050

Table 4.2. Range (mean ± std) values of the log energy and sure entropy features for the VMD modes.

Mode no.	Signals	Log energy entropy Mean ± std	Sure entropy Mean ± std
Mode 1	Myopathy	$-7.70 \times 10^{+03} \pm 491.132$	$1.0040 \times 10^{+03} \pm 0.907$
	Neuropathy	$-6.49 \times 10^{+03} \pm 1.7 \times 10^{+03}$	$1.0659 \times 10^{+03} \pm 137.723$
	Normal	$-7.29 \times 10^{+04} \pm 86.216$	$1.0038 \times 10^{+03} \pm 3.2065$
Mode 2	Myopathy	$-9.38 \times 10^{+03} \pm 660.680$	$1.0018 \times 10^{+03} \pm 0.3259$
	Neuropathy	$-9.67 \times 10^{+03} \pm 2.4 \times 10^{+03}$	$1.0264 \times 10^{+03} \pm 29.755$
	Normal	$-9.877 \times 10^{+03} \pm 549.772$	$1.0011 \times 10^{+03} \pm 1.2009$
Mode 3	Myopathy	$-1.11 \times 10^{+04} \pm 534.318$	$1.0007 \times 10^{+03} \pm 0.1367$
	Neuropathy	$-1.21 \times 10^{+04} \pm 1.5 \times 10^{+03}$	$1.0128 \times 10^{+03} \pm 12.0096$
	Normal	$-1.221 \times 10^{+04} \pm 642.39$	$1.0004 \times 10^{+03} \pm 0.2462$
Mode 4	Myopathy	$-1.1 \times 10^{+04} \pm 437.723$	$1.0006 \times 10^{+03} \pm 0.1146$
	Neuropathy	$-1.14 \times 10^{+04} \pm 2.1 \times 10^{+03}$	$1.0103 \times 10^{+03} \pm 10.367$
	Normal	$-1.332 \times 10^{+04} \pm 247.38$	$1.0001 \times 10^{+03} \pm 0.0501$

In tables 4.1 and 4.2 neuropathy signals have higher values of norm entropy and sure entropy features compared to the other EMG signals. In table 4.2 the myopathy EMG signals have higher values of log energy entropy features when compared to the other EMG signals except for mode 1. In table 4.1 normal EMG signals have a higher Shannon entropy feature compared to other EMG signals except for mode 1. These entropy features determine the degree of uncertainty present in the signal. This uncertainty will give the predictability of a signal. This means that higher entropy values show a less predictable nature for a signal due to the presence of more randomness in a signal. Lower entropy values show high predictability of a signal due to less uncertainty in the resulting signal.

The extracted features are the Shannon ($E1$), log energy ($E2$), norm ($E3$), and sure ($E4$) entropy features. These are given as input to the RF and WKNN classifiers for classifying myopathy, neuropathy, and normal EMG signals. In these classifiers, a ten-fold cross-validation technique is used that evaluates the classification performance. Excepting the test data, the remaining data from the dataset are used for the classifier training. In the RF classifier, the final decision about a class is based on the decision trees. In the WKNN classifier, the classification result of any query depends on the weight assigned to the neighbour. The classification performance of the RF and WKNN classifiers in terms of accuracy (ACC) is presented in tables 4.3 and 4.4. The overall classification performance of the RF and WKNN classifiers based on accuracy are 98.99% and 99.99%, respectively. The confusion matrices of the respective MYO, NEURO, and NOR EMG signals of the distance-weighted k-nearest neighbour and random forest classifier are shown in tables 4.3 and 4.4.

Table 4.3. Confusion matrix of MYO, NEURO, and NOR EMG signals for all modes by the WKNN classifier.

Signal	MYO	NEURO	NOR	Accuracy (%)
MYO	110	1	0	99.099
NEURO	0	148	0	100
NOR	0	0	51	100
Overall classification accuracy (%)				**99.69**

Table 4.4. Confusion matrix of MYO, NEURO, and NOR EMG signals for all modes by the RF classifier.

Signal	MYO	NEURO	NOR	Accuracy (%)
MYO	100	0	1	99.099
NEURO	0	147	1	99.324
NOR	0	1	50	98.039
Overall classification accuracy (%)				**98.99**

Table 4.5. Performance comparison of classification between MYO, NEURO, and NOR EMG signal with the same dataset.

Paper	Classifier	Accuracy (%)
Artameeyanant *et al* [18, 19]	MLPNN	94.75
	SVM	99.07
Proposed work	Random forest (RF)	98.99
	WKNN	99.69

They show that the rate of detection of a neuropathy signal is higher than the myopathy and normal EMG signals. In contrast, the overall classification accuracy of the MYO-NEURO-NOR is 98.99% and 99.99% for all modes by the RF and WKNN classifiers. The proposed framework is compared with other existing techniques on the same dataset, as shown in table 4.5. It has been observed that the obtained VMD modes and their extracted entropy features are helpful for the classification of abnormal and NOR EMG signals. This result clearly shows that the WKNN classifier results an improved classification performance for EMG signals.

4.4 Conclusion

A signal decomposition method VMD is used to decompose an EMG signal into modes. Further, effective entropy-based features are extracted from the computed VMD modes, such as Shannon entropy ($E1$), log energy entropy ($E2$), norm entropy ($E3$), and sure entropy ($E4$). The extracted features are fewer in number, reducing the framework's complexity. These features are fed as input to the RF and WKNN classifier to classify EMG signals. The accuracies for classifying myopathy, neuropathy, and normal EMG classes are 98.99% and 99.69% using the RF and WKNN classifiers, respectively. The proposed framework has better classification results than the existing methods. It can be helpful in the diagnosis of nerve and muscle disorders in clinical applications.

References

[1] Nagineni S, Taran S and Bajaj V 2018 Features based on variational mode decomposition for identification of neuromuscular disorder using EMG signals *Health Inform. Sci. Syst.* **6** 1–10
[2] Khan S M, Khan A A and Farooq O 2019 Selection of features and classifiers for EMG-EEG-based upper limb assistive devices—a review *IEEE Rev. Biomed. Eng.* **13** 248–60
[3] NIH Peripheral neuropathy https://ninds.nih.gov/health-information/disorders/peripheral-neuropathy
[4] Ko K D, El-Ghazawi T, Kim D and Morizono H 2014 Predicting the severity of motor neuron disease progression using electronic health record data with a cloud computing Big Data approach *IEEE Conf. on Computational Intelligence in Bioinformatics and Computational Biology* pp 1–6

[5] Sukumar N, Taran S and Bajaj V 2018 Physical actions classification of surface EMG signals using VMD *In* 2018 Int. Conf. on Communication and Signal Processing (ICCSP) pp 705–9

[6] Demir F, Bajaj V, Ince M C, Taran S and Şengür A 2019 Surface EMG signals and deep transfer learning-based physical action classification *Neural Comput. Appl.* **31** 8455–62

[7] Chada S, Taran S and Bajaj V 2020 An efficient approach for physical actions classification using surface EMG signals *Health Inform. Sci. Syst.* **8** 1–7

[8] Sravani C, Bajaj V, Taran S and Sengur A 2020 Flexible analytic wavelet transform based features for physical action identification using sEMG signals *IRBM* **41** 18–22

[9] Goen A 2014 Classification of EMG signals for assessment of neuromuscular disorders *Int. J. Electron. Electr. Eng.* **2** 242–8

[10] Acharya U R, Ng E Y, Swapna G and Michelle Y S Classification of normal, neuropathic, and myopathic electromyograph signals using nonlinear dynamics method *J. Med. Imaging Health Inform.* **1** 375–80

[11] Bajaj V and Kumar A 2015 Features based on intrinsic mode functions for classification of EMG signals *Int. J. Biomed. Eng. Technol.* **18** 156–67

[12] Gokgoz E and Subasi A 2015 Comparison of decision tree algorithms for EMG signal classification using DWT *Biomed. Signal Process. Control* **18** 138–44

[13] Kehri V, Ingle R, Awale R and Oimbe S 2016 Techniques of EMG signal analysis and classification of neuromuscular diseases *Int. Conf. on Communication and Signal Processing (ICCASP 2016)* pp 481–7

[14] Istenič R, Kaplanis P A, Pattichis C S and Zazula D 2010 Multiscale entropy-based approach to automated surface EMG classification of neuromuscular disorders *Med. Biol. Eng. Comput.* **48** 773–81

[15] Subasi A 2012 Classification of EMG signals using combined features and soft computing techniques *Appl. Soft Comput.* **12** 2188–98

[16] Subasi A, Yilmaz M and Ozcalik H R 2006 Classification of EMG signals using wavelet neural network *J. Neurosci. Methods* **156** 360–7

[17] Güler N F and Koçer S 2005 Classification of EMG signals using PCA and FFT *J. Med. Syst.* **29** 241–50

[18] Artameeyanant P, Sultornsanee S, Chamnongthai K and Higuchi K 2014 Classification of electromyogram using vertical visibility algorithm with support vector machine *Signal and Information Processing Association Annual Summit and Conf. (APSIPA)* pp 1–5

[19] Artameeyanant P, Sultornsanee S and Chamnongthai K 2015 Classification of electromyogram using weight visibility algorithm with multilayer perceptron neural network *7th Int. Conf. on Knowledge and Smart Technology (KST)* pp 190–4

[20] Katsis C D, Exarchos T P, Papaloukas C, Goletsis Y, Fotiadis D I and Sarmas I 2007 A two-stage method for MUAP classification based on EMG decomposition *Comput. Biol. Med.* **37** 1232–40

[21] Torres-Castillo J R, López-López C O and Padilla-Castañeda M A 2022 Neuromuscular disorders detection through time–frequency analysis and classification of multi-muscular EMG signals using Hilbert–Huang transform *Biomed. Signal Process. Control* **71** 103037

[22] Samanta K, Chatterjee S and Bose R 2022 Neuromuscular disease detection based on feature extraction from time–frequency images of EMG signals employing robust hyperbolic Stockwell transform *Int. J. Imaging Syst. Technol.* **32** 1251–62

[23] Tannemaat M R, Kefalas M, Geraedts V J, Remijn-Nelissen L, Verschuuren A J, Koch M, Kononova A V, Wang H and Bäck T H 2022 Distinguishing normal, neuropathic and myopathic EMG with an automated machine learning approach *Clin. Neurophysiol.*

[24] Kiran P U, Abhiram N, Taran S and Bajaj V 2018 TQWT based features for classification of ALS and healthy EMG signals *Am. J. Comput. Sci. Inf. Technol.* **6** 19

[25] Artameeyanant P, Sultornsanee S and Chamnongthai K 2016 An EMG-based feature extraction method using a normalized weight vertical visibility algorithm for myopathy and neuropathy detection *SpringerPlus* **5** 1–26

[26] sein Mousavi S A, Hasan M A, Abdulrazzaq M H and Naghavizadeh M 2020 Diagnosis of myopathy, neuropathy using electromyogram signal and wavelet coefficients *4th Int. Symp. on Multidisciplinary Studies and Innovative Technologies (ISMSIT)* pp 1–3

[27] Belkhou A, Achmamad A and Jbari A 2019 Myopathy detection and classification based on the continuous wavelet transform *J. Commun. Softw. Syst.* **15** 336–42

[28] Goldberger A L, Amaral L A, Glass L, Hausdorff J M, Ivanov P C, Mark R G, Mietus J E, Moody G B, Peng C K and Stanley H E 2000 PhysioBank, PhysioToolkit, and PhysioNet: components of a new research resource for complex physiologic signals *Circulation* **101** e215–20

[29] Dragomiretskiy K and Zosso D 2013 Variational mode decomposition *IEEE Trans. Signal Process.* **62** 531–44

[30] Taran S and Bajaj V 2018 Clustering variational mode decomposition for identification of focal EEG signals *IEEE Sens. Lett.* **2** 1–4

[31] Zavar M, Rahati S, Akbarzadeh-T M R and Ghasemifard H 2011 Evolutionary model selection in a wavelet-based support vector machine for automated seizure detection *Expert Syst. Appl.* **38** 10751–8

[32] Han J, Dong F and Xu Y Y 2009 Entropy feature extraction on flow pattern of gas/liquid two-phase flow based on cross-section measurement *J. Phys.: Conf. Ser.* **147** 012041

[33] Kardam V S, Taran S and Pandey A 2023 Motor imagery tasks based electroencephalogram signals classification using data-driven features *Neurosci. Inform.* **3** 100128

[34] Breiman L 2001 Random forests *Mach. Learn.* **45** 5–32

[35] Han J, Kamber M and Pei J 2011 *Data Mining: Concepts and Techniques* (Waltham)

[36] Cover T and Hart P 1967 Nearest neighbor pattern classification *IEEE Trans. Inf. Theory* **13** 21–7

[37] Chaudhary S, Taran S, Bajaj V and Siuly S 2020 A flexible analytic wavelet transform based approach for motor-imagery tasks classification in BCI applications *Comput. Methods Programs Biomed.* **187** 105325

[38] Dudani S A 1976 The distance-weighted k-nearest-neighbor rule *IEEE Trans. Syst. Man Cybern.* **SMC-6** 325–7

[39] Bajaj V, Taran S and Sengur A 2018 Emotion classification using flexible analytic wavelet transform for electroencephalogram signals *Health Inform. Sci. Syst.* **6** 1–7

IOP Publishing

Data Analytics for Intelligent Systems
Techniques and solutions
Sachin Taran, Chhavi Dhiman and Manjeet Kumar

Chapter 5

Epileptic EEG signal classification using wavelet transform and SVM

Virender Kumar Mehla and Seema Mehla

Automated epileptical seizure identification utilizing electroencephalograms (EEGs) has been a prime area of extensive research. The different frequency bands that make up the EEG signal represent the various emotional and mental processes of humans. The majority of research examines the entire frequency band to find seizures. For the first time, an automated solution using machine learning and only the alpha band is proposed in this study. Seizures are regularly experienced by people with epilepsy, which lowers their quality of life. It is possible to analyse prior seizures and predict upcoming ones when electroencephalograph recordings of these people are carefully divided into seizure-free and seizure-based segments. Discriminative properties can be extracted from an electroencephalogram signal with the aid of modeling. In this work, an automated solution based on wavelet transform and machine learning is proposed for discrimination between normal electroencephalogram signals and signals recorded from epileptic patients. The detection performance of the proposed study has been analysed using many machine learning algorithms and a ten-fold cross-validation method. For the majority of the experiments conducted, the random forest classifier provides the best result among various classifiers when applied to the Neurology and Sleep Centre EEG database, New Delhi.

5.1 Introduction

Epilepsy affects a significantly substantial portion of the global population, with emerging nations seeing the greatest levels of suffering [1]. Epilepsy-related seizures are a neurological condition in which a person loses control of their body, which can result in significant injuries at work or accidents while driving, among other things. Death could ensue from these injuries. The most vulnerable populations are those who have concurrent conditions, older adults, and youngsters, whose prevalence rates vary from 0.7% to 1.0% [2]. For those who have epilepsy, discrimination, misunderstanding, and depression are regular occurrences. In addition, this condition is risky because it can result in mortality if a person with epilepsy partakes in a risky activity [3, 4]. Since these epileptical episodes are the result of brain defects, epilepsy must be identified early so that appropriate treatment can be given. EEG waveforms are frequently employed in the diagnosis of epilepsy because of it is a non-invasive method, and has low cost and excellent temporal resolution. A proper number of electrodes are applied to the scalp at predetermined sites to collect data from patients in the form of electroencephalogram signal waveforms. For the implantation of electrodes on a human scalp, international standards 10–20 have been developed. The electrical activity of brain neurons can be measured with the use of these devices. Simply observing the electroencephalogram data collected from epileptic patients makes it difficult to identify epileptical episodes. A skilled neurologist is needed to correctly point out epileptic seizures from the electroencephalogram waveforms. The necessity for automated epileptical seizure detection, which can help a doctor diagnose a patient, exists despite the likelihood of human error, the scope of incorrect prediction, etc. Epilepsy diagnosis and treatment decisions are both customized. If a person experiences an epileptic seizure and further seizures during the next 24 h, neurologists will diagnose them with epilepsy. If there is a possibility that a patient will have another seizure after two unprovoked ones within the following 8–10 years, they may have epilepsy [5].

If a seizure is recognized early enough, seizure prevention can be implemented to prevent brain damage. The form, amplitudes, and frequency of the brain signal during an epileptic seizure are different from those of regular brain activity [6]. The likelihood that two-thirds of people with epilepsy will have seizure-free lives depends on how quickly the condition is diagnosed and treated with medication and surgery. To identify seizures precisely, researchers have employed a number of neuroimaging techniques [7]. One of the most important instruments in neuroscience for evaluating brain disorders is the electroencephalogram, primarily for seizure detection [8–10]. Neurologists with the necessary knowledge and training use continuous monitoring and electroencephalogram recording interpretation to obtain an accurate epileptic seizure diagnosis. This is a time-consuming, expensive, and intricate task, and because trained people are overworked, it may lead to inaccurate diagnosis. Therefore, various attempts have been undertaken by researchers to automatically detect epileptic episodes.

Sameer et al in [11] utilized a short-time Fourier transform (STFT) to convert non-stationary EEG signals into the time–frequency plane where four statistical features have been identified from the delta frequency band of the electroencephalogram signal. Various classifiers have been used to check the performance of the proposed method in terms of classification accuracy and receiver operating characteristics. Using a random forest (RF) classifier, the proposed work showed 97% accuracy for binary classification problems. Darjani et al in [12] considered epileptic seizure density as one of the useful features that is passed to the k-nearest neighbor (kNN) classifer to check the efficacy of the proposed study. Their study showed a good classification accuracy of 99% during classification between pre-ictal and ictal electroencephalogram signals when implemented and investigated on neurology and sleep electroencephalogram datasets. The authors in [13] developed a hybrid approach in order to investigate high-resolution time–frequency estimation and evaluate the dynamic behavior of electroencephalogram signals. Sharma *et al* [14] utilized time–frequency localized wavelet based features using an orthogonal wavelet filter bank (OWFB) for automatic detection of epileptic seizure detection while investigating the Bonn EEG database and the Neurology and Sleep Centre EEG database. With ten-fold cross-validation, the proposed technique achieved a classification accuracy of 98% for identifying pre-ictal and ictal electroencephalogram signals and a classification accuracy of 100% for identifying inter-ictal and ictal electroencephalogram signals. Kiymik *et al* [15] applied two spectral analysis techniques, namely short-time Fourier transform and continuous wavelet transform (CWT), to detect brain abnormalities using EEG signals. The short-time Fourier transform approach was found to be more appropriate in comparison to the CWT method as it had a shorter processing time. The continuous wavelet transform, on the other hand, took more time, but it recognized the electroencephalogram signals with more accuracy. Wavelet analysis is particularly helpful for the detection of epileptic seizures and the classification of electroencephalogram data since it allows for better results than the short-time Fourier transform when the wavelet analysis parameters are adjusted. Tajmirriahi *et al* [16] applied statistical modeling based on stochastic differential equations for the classification task between seizure and healthy electroencephalogram signals. Using a support vector machine, the proposed study acquired 99.1% accuracy for a binary class problem using the Neurology and Sleep Center database. Five adaptive decomposition strategies for the study of non-stationary and non-linear electroencephalogram signals were described by Carvalho *et al* [17]. It was discovered that, when comparing these methods' classification accuracy, variational mode decomposition and empirical mode decomposition-based approaches performed well. The authors in [18] extracted both temporal and spectral features using discrete wavelet transform (DWT) for the EEG signals analysis. Two non-linear features have been considered as better measures for non-linear electroencephalogram signals. Hadiyoso *et al* [19] applied the wavelet transform method to decompose three classes of electroencephalogram signals into five frequency bands from where two features, namely relative wavelet energy and entropy, were extracted and passed to support vector machine (SVM) classifiers for automatic detection of epileptic seizure. In the ictal versus inter-ictal classification problem, the proposed work achieved an accuracy of 96%. Gupta *et al* in [20] applied a filter bank approach to decompose the electroencephalogram signals into five brain rhythms. For the binary classification of electroencephalogram segments, the Hurst exponent and autoregressive moving average are considered significant features obtained from five brain rhythms and passed as inputs to the support vector machine classifier. The performance of the present study is checked on two publicly available electroencephalogram databases. Sharmila *et al* [21] presented a wavelet based approach for the automatic detection of epileptic seizures where statistical features from wavelet coefficients have been extracted and passed to naïve Bayes and k-nearest neighbor classifiers to check the performance of the proposed study. The authors [22] extracted two features, namely error energy and signal energy, using a fractional linear prediction approach and fed these features to an SVM classifier to classify between ictal and seizure-free signals. The authors in [23] found that permutation entropy is an appropriate feature for the discrimination of binary class problems.

This chapter is organized as follows. The database used in this investigation, the decomposition of the electroencephalogram signals using wavelets, the feature extraction, and the classifiers used are all detailed in sections 5.2 and 5.3, respectively. The simulation results and discussion are presented in section 5.4. The conclusion appears in the last section.

5.2 Proposed method

Ten epileptic patients' segmented electroencephalogram recordings from the Neurology and Sleep Centre in Hauz Khas, New Delhi are included in this study's dataset [24]. The signals were recorded using Grass Tele-factor's Comet AS40 amplification system. Gold-plated scalp electrodes were positioned in accordance with the 10–20 electrode placement scheme when acquiring the electroencephalogram recordings. All the recordings were captured at 200 Hz and then filtered with a band pass filter having cut-off frequencies of 0.5 Hz and 70 Hz, respectively. Pre-ictal, inter-ictal, and ictal are the three categories into which each electroencephalogram segment is segmented. There are 150 segments in the overall dataset (50 for each category). The length of each time series electroencephalogram recording is 5.12 s. Folders named ictal, inter-ictal, and pre-ictal are used to organize the fifty MAT files that make up this dataset.

The pre-ictal folder is designated as set A, the inter-ictal folder is designated as set B, and the ictal folder is designated as set C in this study. These sets are shown in figure 5.1. A flow chart of the proposed method is shown in figure 5.2.

5.2.1 Wavelet transform

One of the most appealing, effective, and promising methods for representing the time–frequency of non-stationary signals, namely electrocardiograms and electro-encephalograms, is the wavelet transform. As it addresses both of their flaws and gives both time and frequency components at any moment, the wavelet transform is primarily applied as a replacement for the Fourier transform and short-time Fourier transform, exhibiting the hidden properties of the original signal more efficiently. In the wavelet transform, high-frequency components can be distinct in the time domain, whereas low-frequency signals can be better resolved in the frequency domain. Wavelet transforms are primarily employed on electroencephalogram signals for preprocessing of the signal, signal decomposition, and feature extraction [25, 26]. The feature of wavelet transform is that its window size can be adjusted by varying the translation and scaling parameters (b, k) which are expressed as

$$\varphi\, b.\, k\, (t) = \frac{1}{\sqrt{k}} \varphi \frac{(t - b)}{k},$$

(5.1)

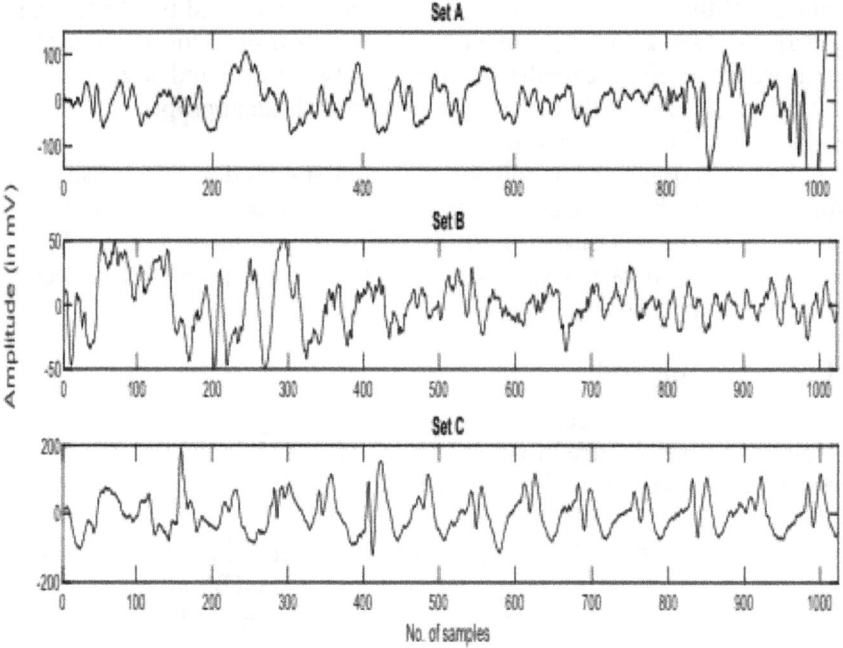

Figure 5.1. A sample of set A, set B, and set C as pre-ictal, inter-ictal, and ictal EEG signals.

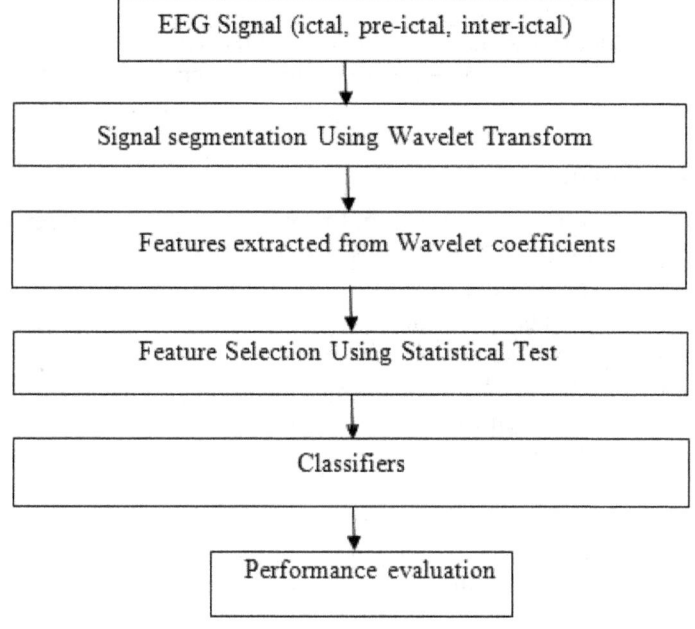

Figure 5.2. A flow chart of the proposed study.

where the function $\psi_{b, k}$, represents the mother wavelet. The term b represents the translation parameter and k represents the scaling factor. The wavelet decomposition method has been employed over the last few decades. In the wavelet transform, the implementation is carried out using a low-pass filter and a high pass filter. In this study the wavelet transform is employed as a filter bank where the original signal is decomposed into various sub-bands using the fifth level of decomposition with Daubechies (db-2) wavelet. Detailed and approximate coefficients have been computed up to the fifth level of decomposition. The detailed coefficients and approximate coefficients represent the high-frequency and low-frequency components of the original signal. In this work, the Daubechies wavelet is employed as a basic wavelet function.

5.2.2 Feature extraction

In this study, seven statistical features, namely mean, maximum value, minimum value, range, mean absolute deviation, standard deviation, and median absolute deviation have been derived from detailed and approximate wavelet coefficients. These wavelet coefficients are extracted from three classes of electroencephalogram signals namely: pre-ictal, inter-ictal, and ictal. These statistical features are discussed in table 5.1.

5.2.3 Feature selection

The feature selection approach plays a significant role in order to acquire a subset of the pertinent characteristics using statistical approaches after computing the statistical features from the wavelet coefficients of each of the three classes of electroencephalogram data. This study employs the MATLAB toolbox to choose relevant characteristics, and the Wilcoxon rank-sum test, with the p-value and z-score at a 95% significance level. By using this method, it is possible to assess how comparable the population locations are (medians equal to zero or not). The alternative hypothesis contends that the medians of the two populations are

Table 5.1. Statistical features computed from wavelet coefficients.

S. No.	Statistical feature	Description
1	Mean	The average value of the original signal.
2	Range	The variation between the maximum and minimum value.
3	Mean absolute deviation	The typical separation between each observation and the data's mean value.
4	Median absolute deviation	A measurement of the signal's variability.
5	Standard deviation	How far data points deviate from the mean value.
6	Maximum	The greatest value in the original signal.
7	Minimum	The lowest value in the original signal.

different, while the null hypothesis contends that their distribution functions are comparable. A relevant feature is one that can be included in the machine learning algorithm with a p-value of less than 0.05 to enhance the classification accuracy. On the other hand, those features with p-values higher than 0.05 are insignificant and can be eliminated.

5.3 Classifications

In this study four classifiers, namely the support vector machine, k-nearest neighbors, decision tree (DT), and ensemble bagged random forest classifiers, are employed to assess the efficacy of the proposed study. A brief description of these classifiers is provided in the following sections.

5.3.1 Support vector machine

One approach that is frequently utilized for classification issues in biomedical signal processing is the support vector machine [27, 28]. The support vector machine was initially used to categorize two groups, and later it was developed to address multiclass classification issues. By maximizing the distance across classes, a support vector machine is utilized to identify the best hyperplane functions [28]. By computing the hyperplane's margin and finding its greatest point, the hyperplane can be identified. A support vector is the nearest pattern. In this study, the proposed approach is validated using both linear and non-linear support vector machines. Cortes and Vapnik developed the support vector machine classifier exclusively using statistical learning theory in 1995. For binary classification, it is among the best supervised classifiers. Additionally, because support vector machine classifiers can avoid overfitting, choose the best kernel, and perform convex optimization, researchers are increasingly employing it to classify patterns. This classifier utilizes the kernel function's architectural design fully. The polynomial kernel, linear kernel, and radial basis functions (RBF) are a few of the well-known support vector machine classifier kernel functions.

5.3.2 k-nearest neighbors

k-nearest neighbors is a straightforward and popular classification technique. This classifier can handle huge and noisy datasets. Because it makes use of local information to forecast unknown data, it is also adaptive. According to Wilson and Martinez [29], it completes the classification assignment using the frequent class of its closest neighbors in the feature space. The k-nearest neighbors method uses a number of distance metrics to specify distance.

In this research, the Euclidean distance is applied based on the training session. An instance is classified by calculating its similarities to its k-nearest neighbors, and the class having the majority of votes is considered the instance's output class.

5.3.3 Decision tree

The supervised machine learning (ML) approach known as the decision tree classifier continuously divides the dataset into smaller and smaller subsets based on a given parameter [30, 31]. Numerous applications, such as financial data analysis, biomedical signal analysis, system control, etc, use this classifier. Regression and classification issues can be resolved using it. The key benefits of this classifier are that it performs well with non-linear requirements and is simple to evaluate and interpret. It also requires less work in preparing the data.

5.3.4 Ensemble bagged random forest

A machine learning approach called random subspace ensemble combines the forecasts from various decision trees that were trained on various subsets of the training dataset's columns [32]. Introducing variety into the ensemble by randomly changing the columns utilized to train each contributing member can improve performance compared to utilizing a single decision tree. It is connected to other decision tree ensembles including random forest, which incorporates concepts from bagging and the random subspace ensemble, and bootstrap aggregation, which builds trees using various samples of rows from the training dataset. The generic random subspace approach can be applied to any machine learning model whose performance meaningfully fluctuates with the selection of input characteristics, despite the fact that decision trees are frequently utilized.

5.4 Result and discussion

In this work, binary classification problems have been considered using three performance metrics, namely sensitivity (Sen), specificity (Spec), and accuracy (Acc), in order to check the performance of the suggested method.

These evaluation parameters are defined as

$$\text{Sen} = \frac{\text{TP}}{\text{TP} + \text{FN}} \times 100\% \tag{5.2}$$

$$\text{Spec} = \frac{\text{TN}}{\text{TN} + \text{FP}} \times 100\% \tag{5.3}$$

$$\text{Acc} = \frac{\text{TP} + \text{TN}}{\text{TP} + \text{TN} + \text{FP} + \text{FN}} \times 100\%. \tag{5.4}$$

TP, TN, FP, and FN, in this context, stand for true-positive rate, true-negative rate, false-positive rate, and false-negative rate, respectively. The major aim of the current study is to discriminate between normal and epileptic signals. The current effort has made use of a publicly available dataset and has taken the categorization difficulty into account. The decomposition of normal and epileptic electroencephalogram signals is achieved by segmenting the entire bandwidth of electroencephalogram signals into wavelet coefficients using discrete wavelet transform. The k-fold cross-validation technique is used to check the success rate of the current study.

The full data sample's ten equally sized segments are each picked as a training set at a certain moment. In the first phase of iteration, the first component is used to test the model, while the subsequent components are for training. The remaining sections are utilized to train the model, while the second component is used for testing in the subsequent iteration. This approach is repeated until a testing set is used for each of the ten portions.

The outcomes of this study are evaluated using pre-ictal, inter-ictal, and ictal signals that are acquired from ten people over a period of 5.12 s and comprised 1024 samples. Figures 5.3–5.5 show the decomposition of three classes of electroencephalogram signals into detailed and approximate coefficients using a wavelet based technique. The detailed and approximate coefficients of pre-ictal, inter-ictal, and ictal signals have been used to extract a total of seven statistical features. For the classification of binary problems, these characteristics are fed to a variety of classifiers including k-nearest neighbor, support vector machine, decision trees, and ensemble bagged random forest classifiers. In tables 5.2 and 5.3 numerous studies that look at the classification of epileptic and regular electroencephalogram signals are contrasted. Comparing the proposed method to other approaches described in the literature, it demonstrated an extraordinary classification accuracy of 100% for differentiating between inter-ictal and ictal electroencephalogram data.

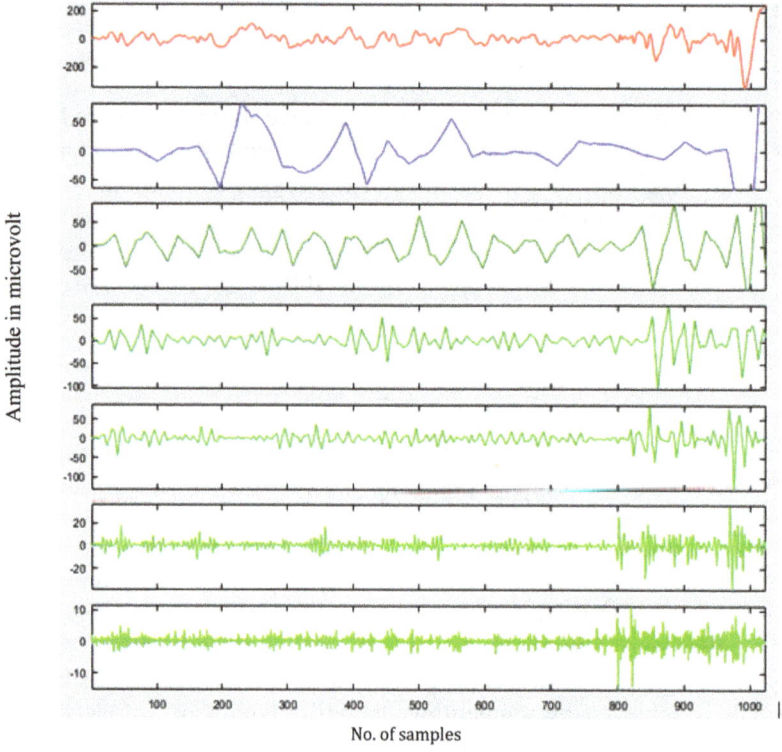

Figure 5.3. Decomposition of pre-ictal EEG signal into wavelet coefficients using Daubechies wavelet.

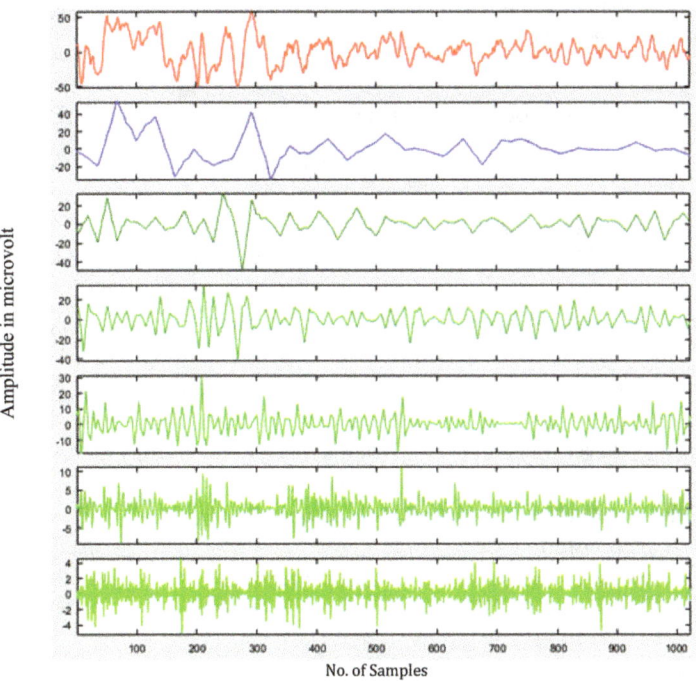

Figure 5.4. Decomposition of inter-ictal EEG signal into wavelet coefficients using Daubechies wavelet.

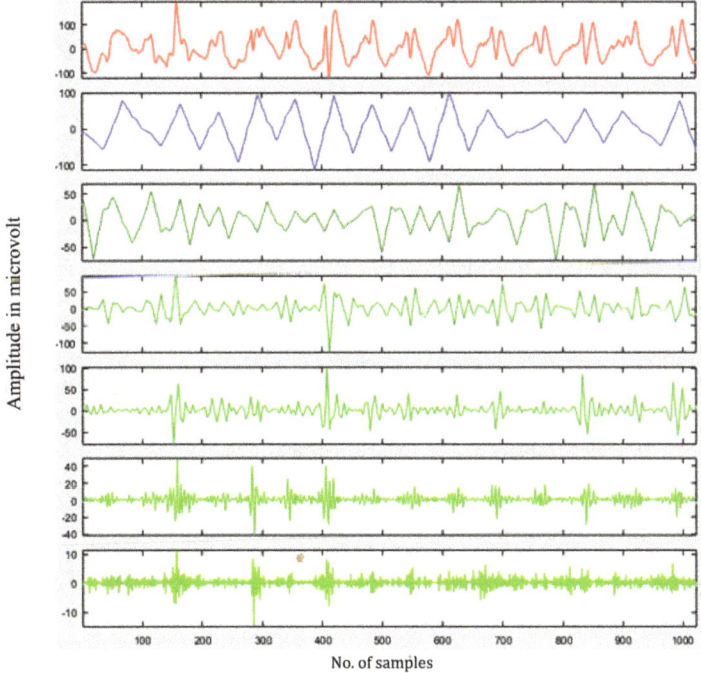

Figure 5.5. Decomposition of ictal EEG signal into wavelet coefficients using Daubechies wavelet.

Table 5.2. A comparison analysis of the proposed study with the existing methods for the discrimination between inter-ictal and ictal EEG signals.

Authors and reference	Technique used	Features used	Evaluation metrics
Sameer *et al* [11]	STFT	Statistical features	Accuracy = 98%
Li *et al* [13]	Hybrid approach	Sub-bands features	Accuracy = 99.30%
Sharma *et al* [14]	OWFB	Entropy-based features	Accuracy = 100%
Tajmirriahi *et al* [16]	Stochastic differential equation	Histogram features	Accuracy = 99.1%
Carvalho *et al* [17]	EMD	Spectral and time domain features	Accuracy = 98.1%
Hadiyoso *et al* [19]	Wavelet transform	Relative energy and entropy-based features	Accuracy = 96%
Gupta *et al* [20]	Discrete cosine transform	Hurst exponent	Accuracy = 97%
Proposed method	Wavelet transform	Statistical features	Accuracy = 100%

Table 5.3. A comparison analysis of the proposed study with the existing methods for the discrimination between pre-ictal and ictal EEG signals.

Authors and reference	Technique used	Features used	Evaluation metrics
Sameer *et al* [11]	STFT	Statistical features	Accuracy = 97%
Li *et al* [13]	Hybrid approach	Sub-bands features	Accuracy = 97.40%
Sharma *et al* [14]	OWFB	Entropy-based features	Accuracy = 98.0%
Tajmirriahi *et al* [16]	Stochastic differential equation	Histogram features	Accuracy = 96.8%
Hadiyoso *et al* [19]	Wavelet transform	Relative energy and entropy-based features	Accuracy = 95%
Gupta *et al* [20]	Discrete cosine transform	Hurst exponent	Accuracy = 79.70%
Proposed method	Wavelet transform	Statistical features	Accuracy = 98.5%

When comparing pre-ictal and ictal signals, the suggested model also demonstrated improved epilepsy detection of 98.5%.

Additionally, figures 5.6–5.9 depict the confusion matrix for the support vector machine, k-nearest neighbors, decision tree, and ensemble bagged random forest

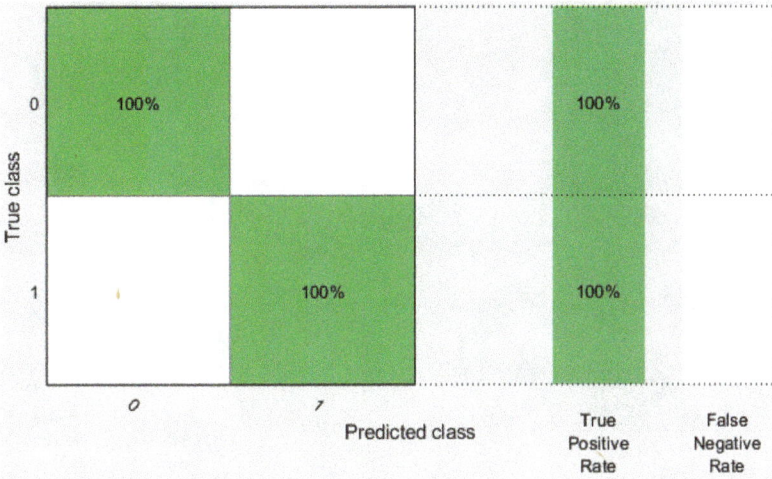

Figure 5.6. Confusion matrix of the SVM classifier to assess the performance in terms of accuracy.

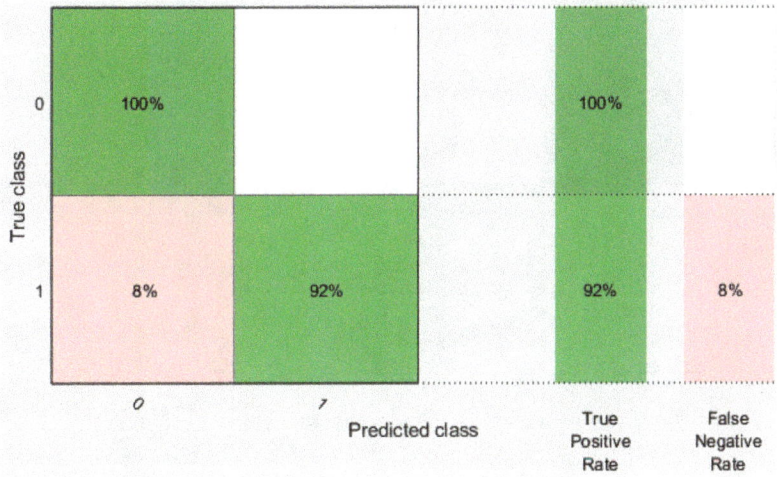

Figure 5.7. Confusion matrix of the kNN classifier to assess the performance in terms of accuracy.

type machine learning algorithms. It is concluded from these figures that the proposed method showed good performance by utilizing a support vector machine and ensemble bagged random forest type supervised machine learning approaches. It is a well-known description used to visualize the classifier's performance to compare predicted and actual classes. It displays the exact number of cases that were classified correctly and inaccurately.

5.5 Conclusion

This work offers a reliable technique for identifying electroencephalogram data as either typical or those captured during an epileptic episode. A total of seven

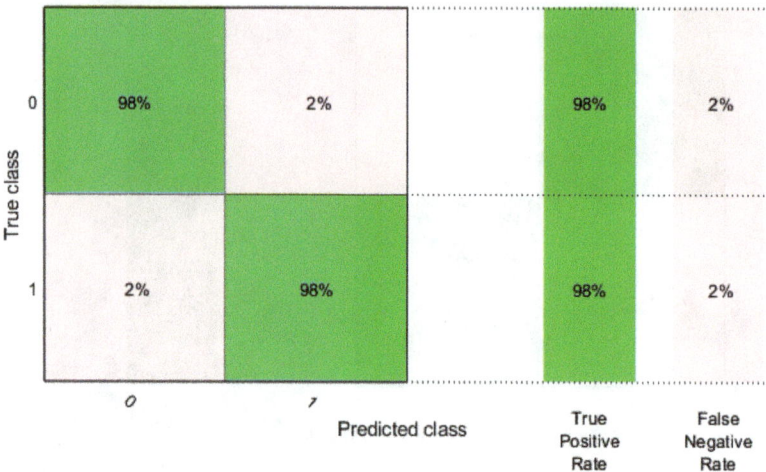

Figure 5.8. Confusion matrix of the decision tree classifier to assess the performance in terms of accuracy.

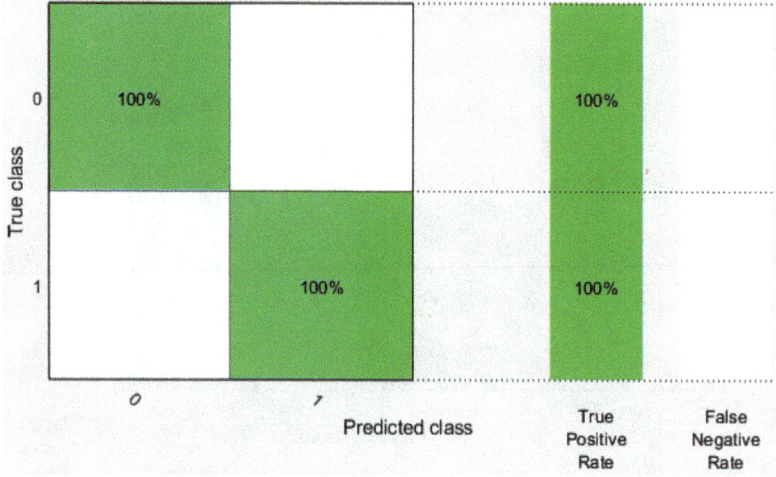

Figure 5.9. Confusion matrix of the ensemble bagged RF classifier to assess the performance in terms of accuracy.

statistical features have been calculated from the wavelet coefficients using the discrete wavelet transform. Application of the Wilcoxon rank-sum test is used to determine the importance of the retrieved features. At the 95% significant threshold, every characteristic with a p-value greater than 0.05 is eliminated. These features are passed to various classifiers such as k-nearest neighbors, support vector machine, decision tree, and ensemble bagged random forest, to assess the precise detection rate of the proposed method. The proposed work, which performed better than prior methods, employed ten-fold cross-validation and showed better results for binary class problems as compared to state-of-the-art methods. Future research could use this method to identify a number of other brain diseases.

References

[1] World Health Organization 2019 Epilepsy https://who.int/en/news-room/factsheets/detail/epilepsy

[2] Kuroda N 2020 Epilepsy and COVID-19: associations and important considerations *Epilepsy Behav.* **108** 107122

[3] Moshé S L, Perucca E, Ryvlin P and Tomson T 2015 Epilepsy: new advances *Lancet* **385** 884–98

[4] Wijayanto I, Rizal A and Humairani A 2019 Seizure detection based on EEG signals using Katz fractal and SVM classifiers *5th Int. Conf. on Science in Information Technology (ICSITech) (Yogyakarta, Indonesia)* pp 78–82

[5] Falco-Walter J J, Scheffer I E and Fisher R S 2018 The new definition and classification of seizures and epilepsy *Epilepsy Res.* **139** 73–9

[6] Tripathi D and Agrawal N 2018 Epileptic seizure detection using empirical mode decomposition based fuzzy entropy and support vector machine *Proc. of the 6th Int. Conf. on Green and Human Information Technology Lecture Notes in Electrical Engineering* S Hwang, S Tan and F Bien p 502

[7] Anyanwu C and Motamedi G K 2018 Diagnosis and surgical treatment of drug-resistant epilepsy *J. Brain Sci.* **8** 1–20

[8] Adeli H, Zhou Z and Dadmehrc N 2003 Analysis of EEG records in an epileptic patient using wavelet transform *J. Neurosci. Methods* **123** 69–87

[9] Liu A, Hahn J S, Heldt G P and Coen R W 1992 Detection of neonatal seizures through computerized EEG analysis *Electroencephalogr. Clin. Neurophysiol.* **82** 30–7

[10] Boashash B, Mesbah M and Colditz P B 2003 Time-Frequency Signal Analysis and Processing: a comprehensive reference (Cambridge, MA: Academic Press) *Elsevier Science* **8** 663–70

[11] Sameer M and Gupta B 2020 Detection of epileptic seizures based on alpha band statistical features *Wirel. Pers. Commun.* **115** 1–16

[12] Darjani N and Omranpour H 2020 Phase space elliptic density feature for epileptic EEG signals classification using metaheuristic optimization method *Knowl.-Based Syst.* **205** 106276

[13] Li Y, Cui W G, Huang H, Guo Y Z, Li K and Tan T 2019 Epileptic seizure detection in EEG signals using sparse multiscale radial basis function networks and the Fisher vector approach *Knowl.-Based Syst.* **164** 96 106

[14] Sharma M, Bhurane A A and Acharya U R 2018 MMSFL-OWFB: a novel class of orthogonal wavelet filters for epileptic seizure detection *Knowl.-Based Syst.* **160** 265–77

[15] Kiymik M K, Guler I, Dizibuyuk A and Akin M 2005 Comparison of STFT and wavelet transform methods in determining epileptic seizure activity in EEG signals for real-time application *Comput. Biol. Med.* **35** 603–16

[16] Tajmirriahi M and Amini Z 2021 Modeling of seizure and seizure-free EEG signals based on stochastic differential equations *Chaos Solitons Fractals* **150** 111104

[17] Carvalho V R, Moraes M F, Braga A P and Mendes E M 2020 Evaluating five different adaptive decomposition methods for EEG signal seizure detection and classification *Biomed. Signal Process. Control* **62** 102073

[18] Gajic D, Djurovic Z, Gligorijevic J, Di G S and Gajic I S 2015 Detection of epileptiform activity in EEG signals based on time-frequency and non-linear analysis *Front. Computat. Neurosci.* **9** 1–16

[19] Hadiyoso S, Irawati I D and Rizal A 2021 Epileptic electroencephalogram classification using relative wavelet sub-band energy and wavelet entropy *Int. J. Eng.* **34** 75–81

[20] Gupta A, Singh P and Karlekar M 2018 A novel signal modeling approach for classification of seizure and seizure-free EEG signals *IEEE Trans. Neural Syst. Rehabil. Eng.* **26** 925–35

[21] Sharmila A and Geethanjali P 2016 DWT based epileptic seizure detection from EEG signals using naïve Bayes/k-NN classifiers *IEEE Access* **4** 7716–27

[22] Joshi V, Pachori R B and Vijesh A 2014 Classification of ictal and seizure-free EEG signals using fractional linear prediction *Biomed. Signal Process. Control* **9** 1–5

[23] Nilocaou N and Georgiou J 2012 Detection of epileptic electroencephalogram based on permutation entropy and support vector machines *Exp. Syst. Appl.* **39** 202–9

[24] Swami P, Panigrahi B K, Nara S, Bhatia M and Gandhi T K 2016 EEG epilepsy datasets

[25] Alyasseri Z A A, Khader A T, Al-Betar M A and Abualigah L M 2017 ECG signal denoising using β-hill climbing algorithm and wavelet transform *8th Int. Conf. on Information Technology* pp 1–7

[26] Kumar H, Pai S P, Vijay G and Rao R 2014 Wavelet transform for bearing condition monitoring and fault diagnosis: a review *Int. J. COMADEM* **17** 9–23

[27] Subasi A and Gursoy I 2010 EEG signal classification using PCA, ICA, LDA and support vector machines *Expert Syst. Appl.* **37** 8659–66

[28] Christianini N and Shawe-Taylor J C 2000 *An Introduction to Support Vector Machines and Other kernel-Based Learning Methods* (Cambridge: Cambridge University Press)

[29] Wilson D R and Martinez T R 2000 Reduction techniques for instance-based learning algorithms *Mach. Learn.* **38** 257–86

[30] Chakure A 2019 *An Introduction to Decision Tree Classifiers* (Toronto: Towards Data Science)

[31] Mehla V K, Kumar A, Singhal A and Singh P 2020 *Noise Removal and Classification of EEG Signals Using the Fourier Decomposition Method, Modeling and Analysis of Active Bio-Potential Signals in Healthcare* (Bristol: IOP Publishing) pp 6.1–6.27

[32] Edla D R, Mangalorekar K, Dhavalikar G and Dodia S 2018 Classification of EEG data from human mental state analysis using random forest classifier *Procedia Comput. Sci.* **132** 1523–32

IOP Publishing

Data Analytics for Intelligent Systems
Techniques and solutions
Sachin Taran, Chhavi Dhiman and Manjeet Kumar

Chapter 6

Imbalanced class problem analysis for lung cancer detection using convolutional neural networks

Om Mishra, Deepak Parashar, Amit Kukker, Aditi Rao, Ananya Srivastava, Anahita, Rajan Mishra and P S Kavimandan

The most common cancer to be diagnosed is lung cancer, especially in men. The early detection of lung cancer allows for effective treatment, perhaps saving lives. The convolution neural network (CNN)-based deep learning methodologies are commonly used to detect lung cancer from CT images. These methodologies result in an imbalanced class distribution. To resolve the problem of an imbalanced class distribution, we propose three models: synthetic minority oversampling technique (SMOTE) analysis, a class-weighted approach, and data augmentation on CNN models. In this chapter different CNN based approaches, ResNet50, AlexNet, and DenseNet-121, are used to classify lung CT images to detect cancer. In this work, the CT image dataset from the Iraq-Oncology Teaching Hospital/National Centre for Cancer Diseases (IQ-OTH/NCCD) is used to evaluate the proposed method. The pre-processing of data is performed and applied to different CNN models. DenseNet-121 achieved an accuracy of 99.4% with SMOTE analysis. The class-weighted and data augmentation approaches achieved accuracies of 97.4% and 95%, respectively, whereas ResNet50 and AlexNet obtained accuracies of 86.05% and 96.5%, respectively. The experimental results show that DenseNet-121 with SMOTE analysis achieved the highest accuracy. The obtained results show that the proposed method outperforms the state-of-the-art approaches.

6.1 Introduction

Around five million deadly cases of lung cancer are recorded per year [1]. When there is an uncontrollable growth in an area of lung tissue, it characterizes a malignant tumor—lung cancer. Early detection of this lethal disease is in great need because early detection helps increase the survival rate of patients and reduces the mortality rate, as the treatment would begin early and is likely to be more curative [2].

These abnormal cancer cells interfere with the regular lung cells' ability to operate, which could result in harmful lung tissue. Tumors occur and form as a result of the expansion of these abnormal processes, which also impairs the lung's ability to provide oxygen to the body through the blood [3].

6.1.1 Problem definition

It is found that CNN based deep learning techniques are widely used to detect lung cancer from CT images. Most of the methodologies used publicly available Kaggle datasets of CT images. These methodologies are less accurate because of the class imbalance distribution issue.

6.1.2 Literature survey and research gap

In the following literature review, recent methodologies to deal with imbalanced class problems are discussed. Among these, an efficient medical screening test that is used for detecting cancer is computed tomography (CT) imaging [4]. The precise information provided by chest CT scan images of aberrant lung cells is utilized to identify lung cancer and reduce the risk of lung cancer death [5]. Recently, deep learning based methods have been used commonly to detect lung cancer from CT images. Gerard *et al* [6] used the FissureNet model for fissure detection in lungs using CT images. The method [7] classified images for lung cancer detection and Covid-19 diagnosis using deep learning. Chen *et al* [8] used a non-multi-model system to detect non-small cell lung cancer. They used the dual path 3D-CenterNET. Masood *et al* [9] proposed a 3D deep CNN for the diagnosis of lung cancer from CT images. Zheng *et al* [10] proposed a CNN based methodology using 2D CT slice images to find nodules in the images. Kumar *et al* [11] improved the multimodality-based system using CNN for PET-CT images. Li *et al* [12] used a generative–discriminative model in a deep neural network to detect lung cancer. Chen *et al* [13] detected lung nodules with the use of LDNNET to classify CT images into cancerous and non-cancerous classes.

6.1.3 Motivation and contribution

Motivated by the above methods, to overcome the gap and limitations of the existing methods, we propose a DenseNet 121-based model with SMOTE analysis for lung cancer detection using CT images. This study employs deep learning techniques to provide approaches for early detection. The proposed work can be used extensively for the efficient and accurate detection of lung cancer using CT Images.

The main contributions of the paper are as follows. (i) To deal with the problem of class imbalance distribution, three approaches (SMOTE analysis, class-weighted, and data augmentation) have been applied. (ii) Three different CNN models, ResNet50, AlexNet, and DenseNet-121, are used to classify lung CT images.

6.2 Methods and materials

6.2.1 Dataset

The Iraq-Oncology Teaching Hospital/National Centre for Cancer Diseases (IQ-OTH/NCCD) lung cancer dataset [14] was obtained through Kaggle (available online at https://www.kaggle.com/datasets/adityamahimkar/iqothnccd-lung-cancer-dataset) and was gathered over three months in the specialized hospital in the autumn of 2019. A sample is shown in figure 6.1. It contains CT scans of both healthy volunteers and patients with lung cancer at various stages of the disease.

Oncologists and radiologists who were present in these two centers annotated the IQ-OTH/NCCD slides. In total 1190 images were taken from slices of CT scans of 110 cases to make up the dataset. These cases are divided into three categories: benign, malignant, and normal cases [15].

The scanner was a Siemens SOMATOM model. The CT procedure calls for a slice thickness of 1 mm, a window width of 350–1200 HU a, and a window center of 50–600 HU a, with a breath-hold throughout the inspiration peak. Before performing the analysis, all scanned images were de-identified. These slices range in number from 80 to 200, wherein each is depicting a CT image of the human chest area from various angles to have a wide view. The 110 instances that we have in our dataset are different in terms of gender, age, educational level, locale, and lifestyle of the people [15]. The system architecture plan is to train the sample image files. First, the data are pre-processed and then fed to the deep learning model. Once the model is trained, it is ready to test sample images and predict their class. The proposed methodology involves the classification of the CT images into three classes: normal, benign, and malignant. In the proposed framework different CNN architectures (DenseNet-121, ResNet50, and AlexNet) are used with SMOTE, class-weighted,

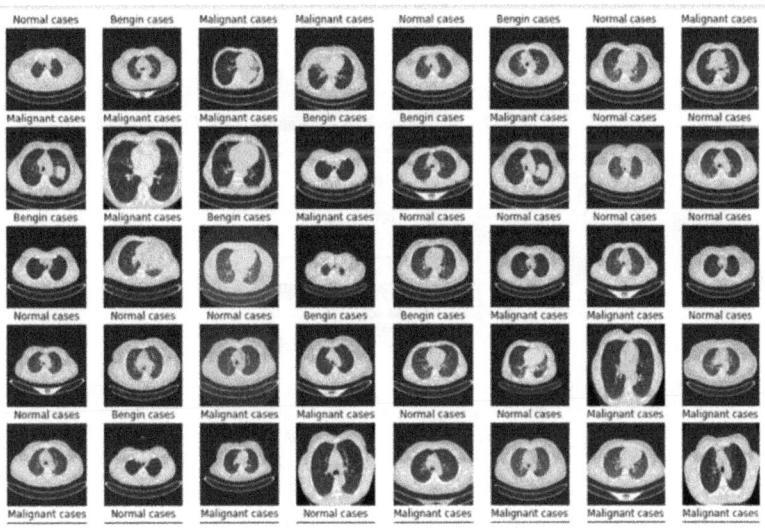

Figure 6.1. CT images of IQ-OTH/NCCDdataset. (Reproduced from [14]. CC0 1.0.)

and data augmentation approaches. The architecture of the proposed framework is given below.

6.2.2 ResNet50 and AlexNet

A well-known neural network called ResNet, which stands for 'residual network', provides the basis for numerous computer vision applications shown in figure 6.2. A 50-layer convolutional neural network is called ResNet50, wherein there are 48 convolutional layers, one max-pool layer, and one average pool layer [16]. Residual neural networks (RNNs) are artificial neural networks (ANNs) that generate networks from residual blocks [17].

Three fully linked layers and five convolutional layers make up the eight layers that make up the architecture of AlexNet, a form of CNN. AlexNet was the initial convolutional network to use a GPU to extract maximum power [18]. The architecture of the AlexNet Model is shown in figure 6.3.

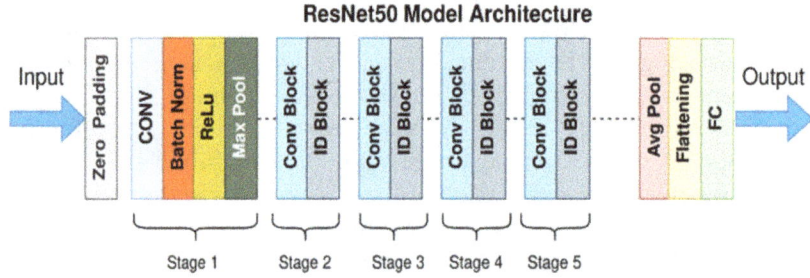

Figure 6.2. The architecture of ResNet50. (Image by Gorlapraveen123, Wikimedia Commons https://commons.wikimedia.org/wiki/File:ResNet50.png. CC BY-SA 4.0.)

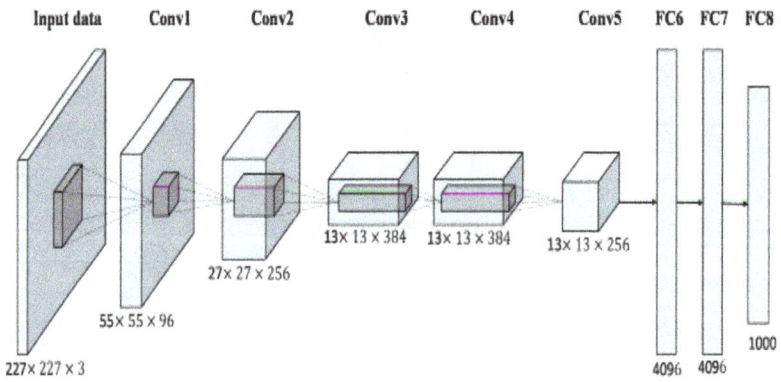

Figure 6.3. The architecture of AlexNet. (Reproduced with permission from [27]. CC BY 4.0.)

6.2.3 DenseNet

In a dense convolutional neural network [19], the feature mapping of all previous layers is done through the concatenation of features. These concatenated features are used as input to the next layer, shown in figure 6.4. If we compare DenseNet to other CNN based models, it requires fewer parameters as redundant parameters are neglected.

In figure 6.4 suppose the DenseNet shown has j number of layers $(X_1, X_2, ..., X_{j-1}, X_j)$ with features $F_1, F_2, ..., F_{j-1}, F_j$, respectively. In the DenseNet model, the input to the jth layer will be the concatenated features of all previous layers $(X_1, X_2, ..., X_{j-1})$. The outcome is then generated by the CNNs using layer-based predictions. Then we would obtain some sort of a vector that would show the likelihood of being in a particular class. To concatenate the features of the previous layers, the dimension of the feature map need to be constant. A constant sampling of the layers is achieved by the use of dense blocks. A convolutional layer has a more intricate structure than a conventional neural layer, which is distinguished by a weight matrix in two dimensions. To filter the original data and extract input features, convolutional layers are used as shown in AlexNet architecture [28] (figure 6.5).

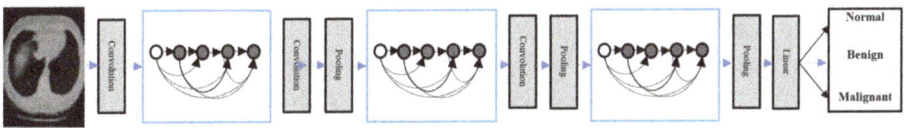

Figure 6.4. The steps involved in DenseNet.

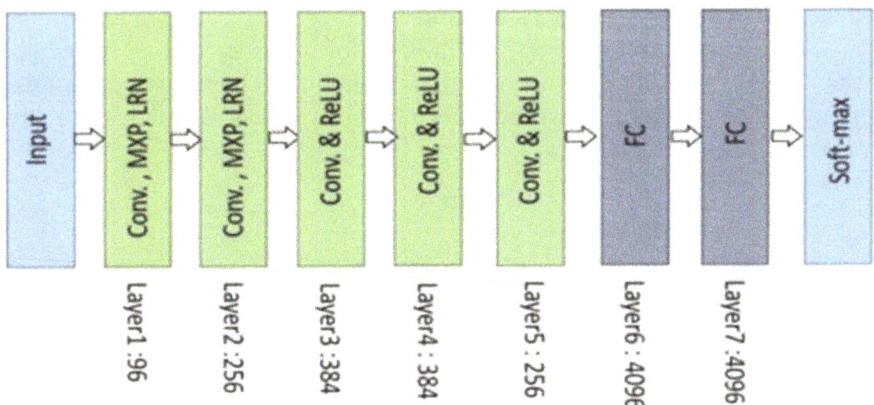

Figure 6.5. Implementation of the AlexNet architecture. (Reproduced from [28]. CC BY 4.0.)

6.2.4 SMOTE analysis to oversample the data

The synthetic minority oversampling technique is referred to as SMOTE [21]. SMOTE is a better approach for resolving classification problems involving imbalanced data. Observed variables are said to be unbalanced when they greatly diverge from the possible values for that variable. There are many observations of one type and a few of another.

To augment data, the SMOTE algorithm adds fictitious data points to the real ones. In general, we can describe SMOTE as a particular kind of data augmentation technique or a better oversampling alternative [22]. Utilizing SMOTE, we can make new data points that slightly vary from the actual data points rather than cloning them.

The SMOTE algorithm works in the following manner. We choose to have a random sample from the minority class. Each observable in this sample must have k closest neighbors, which we could define. We evaluate the direction of the connection between the current data point and one of those neighbors using one of those neighbors. We enhance the vector by a random number between 0 and 1. To create the synthetic data point, this is combined with the original data point. We built our DenseNet model after doing the SMOTE analysis.

6.2.5 Class-weighted approach

The bulk of machine learning approaches is not particularly useful with skewed class data. Skewed class distribution in data can be addressed by giving the majority and minority classes different weights [23]. The weight disparities will have an impact on the classification of the classes during the training phase. By increasing the class weight for the minority class while concurrently reducing or degrading the class weight for the majority voting class, we can say the entire scheme is intended to punish the minority class for causing the misclassification and affecting the results significantly.

The majority of scikit-learn classifier modeling libraries, as well as some boosting-based libraries, including LightGBM and Catboost, provide an internal setting called 'class weight' that enables us to optimize the scoring for the minority class by what we have learned so far. Class weight is set to 'None' by default, which means that the weights for both classes are equal. In addition, we have two options: we can either mark it as 'balanced' or we can pass a dictionary that has manual weights for both classes.

6.2.6 Data augmentation

We employ the methodology of data augmentation when we want to increase the amount of data by literally increasing the number of images that already exist [24]. Thus new artificial data are generated by making copies of the original data that we have modified. These methods will be helpful to make our model more generalized towards different angles of image data, thereby helping in reducing the overfitting of the trained model. Hence, data augmentation methods result in regulating the performance of the model by oversampling the data samples.

6.3 Results and discussion

The CNN was constructed for interpreting lung CT for cancer which is a challenging problem. In this research, we experiment with utilizing the deep neural network methods [26], namely DenseNet-121, ResNet50, and AlexNet, to extract high-level features from raw data, perform proper image classification, and show that they are effective for lung cancer detection.

We first pre-process the data and convert them into the model-required format. Then, we remove the issue of imbalanced data distribution using SMOTE analysis, a class-weighted approach, or data augmentation. We then experiment to determine how the suggested solution successfully learns to correctly classify the lung CT scanned images as normal, benign, or malignant cases, then reports are accordingly generated for the lung cancer patients.

6.3.1 Implementation of the DenseNet-121 model

Table 6.1 shows the details of the different DenseNet convolutional neural networks [20]. DenseNet-121 has various dense blocks. Each dense block has various numbers of layers. There are two convolution, 1×1 and 3×3, layers. The 3×3 kernel is used for the convolution operation and 1×1 is used as the bottleneck layer. Transition layers are present between two dense blocks. In each transition layer, one 1×1 convolution layer and one 2×2 pooling layer are present. To categorize lung CT scan images as malignant, benign, or normal instances, we employ a convolutional neural network (CNN) model. The architecture is made up of activation functions,

Table 6.1. Various components of the architecture of DenseNet-121.

Layers	Output size	DenseNet-121
Convolution	112×112	7×7 convolution with stride 2
Pooling	56×56	3×3 max-pool, with stride 2
Dense block 1	56×56	$\begin{bmatrix} 1 \times 1 \text{convolution} \\ 3 \times 3 \text{convolution} \end{bmatrix} \times 6$
Transition layer 1	56×56 (convolution), 28×28 (pooling)	1×1 convolution, 2×2 average pool with 2 stride
Dense block 2	28×28	$\begin{bmatrix} 1 \times 1 \text{convolution} \\ 3 \times 3 \text{convolution} \end{bmatrix} \times 12$
Transition layer 2	28×28 (convolution), 14×14 (pooling)	1×1 convolution, 2×2 average pool with 2 stride
Dense block 3	14×14	$\begin{bmatrix} 1 \times 1 \text{convolution} \\ 3 \times 3 \text{convolution} \end{bmatrix} \times 24$
Transition layer 3	14×14 (convolution), 7×7 (pooling)	1×1 convolution, 2×2 average pool with 2 stride
Dense block 4	7×7	$\begin{bmatrix} 1 \times 1 \text{convolution} \\ 3 \times 3 \text{convolution} \end{bmatrix} \times 16$
Classifying layer	1×1 (classifying layer)	7×7 global average pool, 1000D fully connected

dropout layers, layers that are fully connected, pooling layers, and convolutional layers. Filters are applied to the input in the convolutional layers.

Downsampling the feature maps created by the convolutional layers is done by the pooling layers. After the number 1 convolutional layer, a max-pooling layer is used. The 'flatten' and 'dense' layers of the fully connected layers extract the properties before giving them to the SoftMax function, which calculates the probability of being classified in each class.

In our dataset, we initially observed that there is variation in the sizes of images that belong to the three classes of malignant, benign, and normal cases. We can see that there are 120 images of benign cases of size 512×512, 501 malignant images of size 512×512, 1 image of size 404×511, 28 images of size 512×801, 31 images of size 512×623, and for normal images there is a count of 415 images of 512×512 and 1 image of size 331×506. To ensure that our model is fed processed data of similar sizes we performed pre-processing and finally resized our images to 256×256. Now, performing the train test split to divide our data into training and validation data, we find that we have 822 images in the training set and 275 in the testing set, every image having the size 256×256. After applying SMOTE, we can see that the number of data points in each class has been equalized and prediction is very accurate.

6.3.2 Implementation of the AlexNet model

The implementation of the AlexNet model, shown in figure 6.5, has a total of eight layers. There are five convolutional and three fully connected layers. The first two convolutional layers are connected to an overlapping max-pooling layer which reduces the dimension of the image and extracts a maximum number of features from the images. The third, fourth, and fifth convolutional layers are then connected directly to the fully connected layer and, in the end, the final layer is connected to the SoftMax layer. When we are working with AlexNet, we can observe from the model summary that the first convolutional layer has 96 distinct receptive filters that are each 11×11 pixels in size to perform convolution and max-pooling using local response normalization which is LR. With only a stride size of two, 3×3 filters were employed to implement max-pooling operations. In second layer, identical operations with first layer are performed. With 384, 384, and 296 feature maps, the third, fourth, and fifth convolutional layers all employ 3×3 filters. It uses two fully connected (FC) layers with dropout and a SoftMax layer at the end. For this model, two parallel networks are trained with identical structures and the same number of feature maps. In this network, the principles of local response normalization (LRN) and dropout are introduced [13].

For AlexNet, we again load our data with oversampling and resize the image to 64×64. The training images that we feed now number 1619, with 695 validation images and 30 images for testing. The model is compiled with the Adam optimizer, the loss is categorical cross entropy as it is a multiclass classification, and the metric used is accuracy. The model was trained on 20 epochs with a batch size of 32. The model was tested on a validation set and a test set. Then we augmented the images and again trained the model on augmented data with 10 epochs, thereby increasing the accuracy of the AlexNet model and comparing the two results obtained.

6.3.3 Implementation of the ResNet50 model

ResNet50 is a deep convolutional neural network with 50 layers, 48 convolutional layers, 1 max-pool layer, and 1 global average pool layer. In our model we have about 64 filters on a 7×7 convolutional layer. An upper pool layer has three convolutional layers—a 1×1 layer with 64 filters, a 3×3 layer with 64 filters, and a 1×1 layer with 256 filters that is repeated three times—and also we can establish that between the layers, there are skip connections as well.

Three convolutional layers, a 1×1 layer with 128 filters, a 3×3 layer with 128 filters, and a 1×1 layer with 512 filters that is repeated four times, and here there are also skip connections between the layers. Consequently, there are 12 layers altogether: three convolutional layers, a 1×1 layer with 256 filters, a 3×3 layer with 256 filters, and a 1×1 layer with 1024 filters that is repeated six times. Consequently, there are 18 layers altogether: three layers of convolutional processing, a 1×1 layer with 512 filters, a 3×3 layer with 512 filters, and a 1×1 layer with 2048 filters that is repeated three times. Between the layers, there are skip connections as well. There are nine levels altogether. The overall number of layers is 50 due to the final average pool layer and a fully linked layer with 1000 nodes and SoftMax activation function.

The model training is generalized by the skip connections, which help lessen overfitting that could develop when utilizing the optimizers. The last fully connected layers were eliminated from our version in favor of a flattened layer, a dense layer with 128 nodes and ReLU activation, a dropout layer with a probability of 0.5, and a final dense layer with three nodes and SoftMax activation. The model is put together using the Adam optimizer, which combines gradient descent with momentum and RMSprop, and easily and quickly reaches the global minima (the learning rate considered is 0.001, $\beta1$ is 0.9, $\beta2$ is 0.99, and decay is 1e−09); the loss is the categorical cross entropy since our motive is multiclass classification, and the metric is accuracy. The model was tested on testing images after being trained on training images for ten epochs with a batch size of 64.

6.3.4 Performance evaluation parameters

Accuracy. The percentage of correctly predicted events that the model produced. How successfully did the model anticipate the right class label? Accuracy gained while testing the model on the validation set of images is known as 'validation accuracy'.

Loss. This represents the degree to which the model did not predict the class label or the mistake. The term 'validation loss' refers to the loss that the model experienced while being tested on the validation set of images.

Confusion matrix. The confusion matrix is a tool for describing how well a trained classifier performs. It shows how many instances were successfully classified (true positive and true negative) and how many were wrongly labeled (false negative and false positive). Table 6.2 gives the comparison of the proposed method with other deep learning models (ResNet50 and AlexNet). To incorporate class imbalance distribution, three analyses, SMOTE, class-weighted, and AlexNet, have been used. Table 6.1 shows that the proposed DenseNet-121 model achieved the highest accuracy compared to the other two models. Figure 6.6(a) shows the model accuracy and model loss concerning the epoch DenseNet-121 based model using SMOTE analysis. In the training phase, the training accuracy is 99.7% and in the validation phase the validation accuracy is 99.4%.

Figure 6.6(b) is for class-weighted analysis. It represents a training accuracy of 99.7% and a validation accuracy of 97.4%. We also performed DenseNet-121 after

Table 6.2. Comparison in terms of model accuracy (%) with other deep learning models for three different analyses for lung cancer detection using the Kaggle database.

Dataset	Method/model	Performance in accuracy (%)		
		SMOTE	Class-weighted	Data augmentation
Kaggle database [14]	ResNet [16]	86.05	82.52	48.49
	AlexNet 50 [18]	96.5	94.33	93.36
	Proposed	**99.4**	97.42	92.21

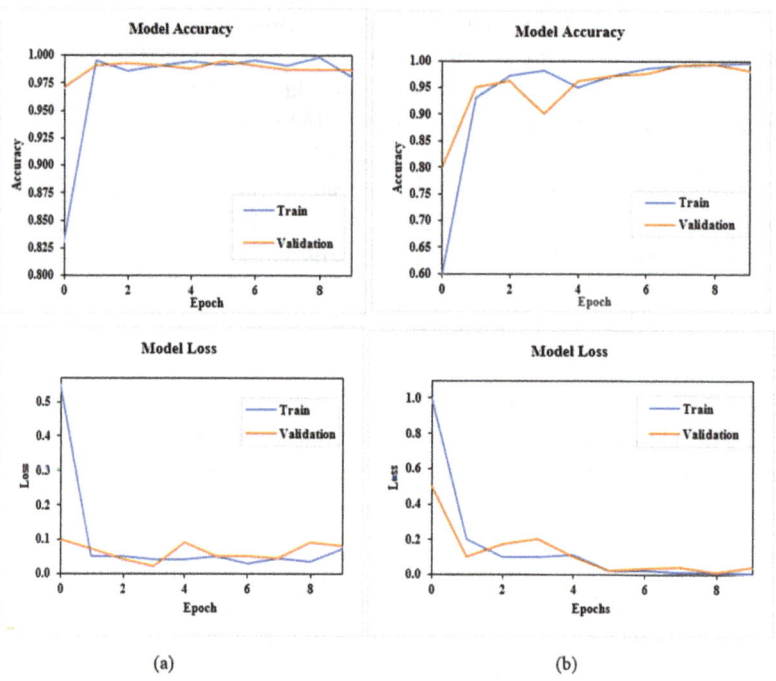

(a)　　　　　　　　　　(b)

Figure 6.6. Plots of model loss and accuracy. (a) DenseNet-121 with SMOTE analysis. (b) Class-weighted approach.

data augmentation and obtained a training accuracy of 94.7% and a validation accuracy of 92.2%. The ResNet50 model for data augmentation is shown in figure 6.7(a). Training accuracy is 99.2% and validation accuracy is 48.49%. It performed poorly on our dataset and the validation accuracy at every epoch was constant. Figure 6.7(b) shows a training accuracy of 98.3% and validation accuracy of 93.36% for the AlexNet model with data augmentation analysis.

The proposed method is also compared with the recent state-of-the-art methodologies in table 6.3. Recently, Li *et al* [12] used the CNN model with a segmentation approach in whole-slide histopathology images, they obtained accuracy, sensitivity, and specificity of 95.05%, 90.52%, and 95.31%, respectively. Chen *et al* [13] used LDNNET for cancer detection in the lung using a dense neural network, they

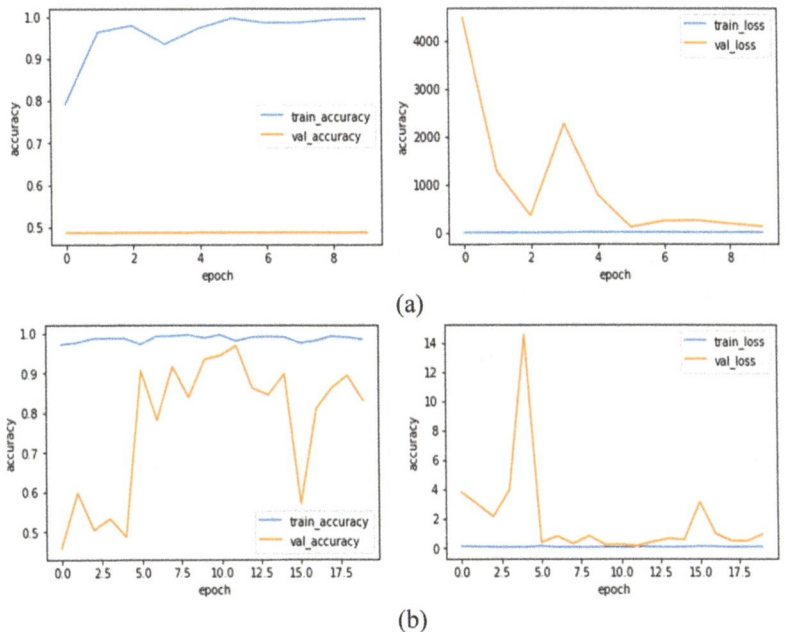

Figure 6.7. (a) ResNet50 metrics. (b) Testing metrics for Alexnet after augmentation.

Table 6.3. Comparison with the existing methods for lung cancer detection using the Kaggle database.

Dataset	Method/model	Ac (%)	Sn (%)	Sp (%)
Kaggle database [14]	Li *et al* [12]	95.05	90.52	95.31
	Chen *et al* [13]	98.83	99.45	98.20
	Alzubaidi *et al* [5]	97	96	97
	Kumar *et al* [11]	99	NA	NA
	Masood *et al* [9]	98.51	98.4	92
	[Proposed method]	**99.4**	100	99.10

reported accuracy, sensitivity, and specificity of 98.83%, 99.45%, and 98.20%, respectively. Alzubaidi *et al* [5] used global and local feature extraction frameworks for lung cancer detection; they achieved accuracy, sensitivity, and specificity of 97%, 96%, and 97%, respectively. Kumar *et al* [11] used co-learning feature fusion maps from PET-CT images, they reported an accuracy of 99%. In another study, Masood *et al* [9] used a cloud-based automated clinical decision support system for the identification of lung cancer; they obtained accuracy, sensitivity, and specificity of 98.51%, 98.40%, and 92%, respectively. In contrast, we obtained far better performance in terms of accuracy, sensitivity, and specificity of 99.4%, 100%, and 99.10%, respectively. The experimental results show that the proposed method outperformed to state-of-the-art-approaches for lung cancer detection using CT images.

6.4 Conclusion

In this paper, we implemented three approaches—SMOTE, class-weighting, and data augmentation analysis—on three different powerful deep learning CNN based techniques—DenseNet-121, ResNet50, and AlexNet—to overcome the problems of class imbalance in CT images for lung cancer detection. The experimental results showed that DenseNet-121 had the maximum accuracy of 99.4% with SMOTE, whereas the other two approaches had accuracies of 97.4% and 92.2% respectively. We used pre-processed data on ResNet50 and AlexNet which showed accuracies of 86.05% and 96.6%, respectively. It can be seen from the proposed work that DenseNet-121 with SMOTE analysis achieved better results compared to the other approaches. In the future, the proposed method could be implemented to detect Covid-19 disease from CT images.

Acknowledgments

The authors wish to acknowledge the assistance and encouragement from Symbiosis International University, Symbiosis Institute of Technology Pune, India.

References

[1] Zhang G 2018 Automatic nodule detection for lung cancer in CT images: a review *Comput. Biol. Med.* **103** 287–300

[2] Jiang J, Hu Y C, Halpenny D, Hellmann M D, Deasy J O, Mageras G and Veeraraghvan H 2019 Multiple resolution residually connected feature streams for automatic lung tumor segmentation from CT images *IEEE Trans. Med. Imaging* **38** 134–44

[3] Balagurunathan Y 2021 Lung nodule malignancy prediction in sequential CT scans: summary of ISBI 2018 challenge *IEEE Trans. Med. Imaging* **40** 3748–61

[4] Nadkarni N S and Borkar S 2019 Detection of lung cancer in CT images using image processing *3rd Int. Conf. on Trends in Electronics and Informatics (ICOEI)* pp 863–6

[5] Alzubaidi M A, Otoom M and Jaradat 2021 Comprehensive and comparative global and local feature extraction framework for lung cancer detection using CT scan images *IEEE Access* **9** 158140–54

[6] Gerard S E *et al* 2019 FissureNet: a deep learning approach for pulmonary fissure detection in CT images *IEEE Trans. Med. Imaging* **38** 156–66

[7] Hişam D and Hişam E 2021 Deep learning models for classifying cancer and COVID-19 lung diseases *Innovations in Intelligent Systems and Applications Conf. (ASYU)* pp 1–4

[8] Chen L *et al* 2022 Multimodality attention-guided 3-D detection of nonsmall cell lung cancer in 18F-FDGPET/CT images *IEEE Trans. Radiat. Plasma Med. Sci.* **6** 421–32

[9] Masood A *et al* 2020 Cloud-based automated clinical decision support system for detection and diagnosis of lung cancer in chest CT *IEEE J. Transl. Eng. Health Med.* **8** 1–13

[10] Zheng S *et al* 2020 Automatic pulmonary nodule detection in CT scans using convolutional neural networks based on maximum intensity projection *IEEE Trans. Med. Imaging* **39** 797–805

[11] Kumar A, Fulham M, Feng D and Kim J 2020 Co-learning feature fusion maps from PET-CT images of lung cancer *IEEE Trans. Med. Imaging* **39** 204–17

[12] Li J, Tao Y and Cai T 2021 Predicting lung cancers using epidemiological data: a generative-discriminative framework *IEEE/CAA J. Autom. Sin.* **8** 1067–78

[13] Chen Y *et al* 2021 LDNNET: towards robust classification of lung nodule and cancer using lung dense neural network *IEEE Access* **9** 50301–20

[14] The IQ-OTH/NCCD lung cancer dataset kaggle.com (https://kaggle.com/hamdallak/the-iqothnccd-lung-cancer-dataset)

[15] Al-Yasriy H F *et al* 2020 Diagnosis of lung cancer based on CT scans using CNN *IOP Conf. Ser.: Mater. Sci. Eng.* **928** 022035

[16] Salama W M, Shokry A and Aly M H 2022 A generalized framework for lung cancer classification based on deep generative models *Multimed Tools Appl.* **81** 32705–22

[17] He T *et al* 2019 Bag of tricks for image classification with convolutional neural networks *IEEE Conf. on Computer Vision and Pattern Recognition* pp 558–67

[18] Xu Y, Wang Y and Razmjooy N 2022 Lung cancer diagnosis in CT images based on AlexNet optimized by modified Bowerbird optimization algorithm *Biomed. Signal Process. Control* **77** 103791

[19] Saini M and Susan S 2022 Diabetic retinopathy screening using deep learning for multi-class imbalanced datasets *Comput. Biol. Med.* **149** 105989

[20] Belciug S 2022 Learning deep neural networks' architectures using differential evolution. Case study: medical imaging processing *Comput. Biol. Med.* **146** 105623

[21] Alves E, Leal A, Lopes M and Fonseca A 2021 Performance analysis among predictive models of lightning occurrence using artificial neural networks and SMOTE *IEEE Lat. Am. Trans.* **19** 755–62

[22] Wei J, Lu Z, Qiu K, Li P and Sun H 2020 Predicting drug risk level from adverse drug reactions using SMOTE and machine learning approaches *IEEE Access* **8** 185761–75

[23] Joshua N *et al* 2021 3D CNN with visual insights for early detection of lung cancer using gradient-weighted class activation *J. Healthcare Eng.* **2021** 6695518

[24] Tang Z, Chen K, Pan M, Wang M and Song Z 2019 An augmentation strategy for medical image processing based on statistical shape model and 3D thin plate spline for deep learning *IEEE Access* **7** 133111–21

[25] Toda R *et al* 2022 Lung cancer CT image generation from a free-form sketch using style-based pix2pix for data augmentation *Sci. Rep.* **12** 12867

[26] Deo S *et al* 2023 Video tampering detection using machine learning and deep learning *Advanced Computing, IACC 2022, Communications in Computer and Information Science* ed D Garg, V A Narayana, P N Suganthan, J Anguera, V K Koppula and S K Gupta (Cham: Springer) vol 1782

[27] Han X, Zhong Y, Cao L and Zhang L 2017 Pre-trained AlexNet architecture with pyramid pooling and supervision for high spatial resolution remote sensing image scene classification *Remote Sens.* **9** 848

[28] Alom M Z *et al* 2019 State-of-the-art survey on deep learning theory and architectures *Electronics* **8** 292

IOP Publishing

Data Analytics for Intelligent Systems
Techniques and solutions
Sachin Taran, Chhavi Dhiman and Manjeet Kumar

Chapter 7

An end-to-end content-aware generative adversarial network based method for multimodal medical image fusion

Manisha Das, Deep Gupta and Ashwini Bakde

The fusion of multimodality medical images combines the most relevant information of the source modalities to improve diagnostic accuracy. Most recently, end-to-end deep learning (DL) based image fusion approaches have shown more accurate and robust fusion performance by combining the feature extraction, selection, and fusion steps together. However, the DL model must learn to extract and preserve distinct features from source images, each of which portrays distinct aspects of the underlying tissues. In this paper, an end-to-end multimodal medical image fusion method is presented using a content-aware generative adversarial network based approach. During the training phase, the generator is trained to generate fake fused images from the source multimodal image pairs. Two discriminators are used to differentiate between the source and fake fused images. Further, the loss function is made content-aware to elevate the preservation of the characteristic information of each of the source images. During the testing phase, the discriminators are discarded and the learned generator model is used to generate the fused images from a multimodal image pair. The subjective and quantitative performance analysis of the extensive experimentation demonstrates that the proposed method achieves notable improvement compared to recent DL based image fusion methods.

7.1 Introduction

Medical image fusion merges images acquired from different sensors to reduce information redundancy and obtain a more comprehensive and concise representation of various tissue structures [1, 2]. It is used extensively by clinicians and radiologists for interpreting more factual and accurate diagnostic inferences of various brain-related anomalies [3, 4]. There exists a wide variety of medical imaging modalities that are used to represent the structures and functions of various tissues, such as magnetic resonance (MR) imaging, computed tomography (CT), single photon emission computed tomography (SPECT), and positron emission tomography (PET) [5]. However, a single modality image does not present comprehensive diagnostic information. The information provided by a pair of multimodal source images is complementary and viewing them in an integrated single fused image can increase the information content which is beneficial for further diagnosis and treatment procedures [6].

There exists a variety of spatial and transform domain image fusion methods. The spatial domain methods process the source pixels directly while in the transform domain the images are first transformed and then processed further [7, 8]. In addition to this, spiking neural network based methods are also used for medical image fusion, wherein they are used to extract and fuse features consistent with human perception [9]. Some hybrid methods based on sparse representation, fuzzy logic, dimensionality reduction, superpixel segmentation, etc, have also been presented to further improve the fusion efficacy [10–13]. The prime limitation of these methods is the dependence of the fusion performance on the quality of extracted features and mostly handcrafted fusion rules [14]. This affects the overall fusion efficacy and also affects the robustness of the method to fuse a variety of multimodal medical images [14, 15].

Recently, DL based fusion approaches based on convolutional neural networks (CNNs) [16, 17], transfer learning models [18], and autoencoders (AEs) [14] have been proposed, demonstrating better fusion performance than the conventional handcrafted fusion methods. More recently, unsupervised DL models have been explored based on the generative adversarial network (GAN) [19] and its variants [20, 21], dense autoencoders [22, 23], etc. However, most of the time these methods lack the ability to preserve the crucial diagnostic information, largely because the designed loss function tends to overlook the content of the source images. Further, most of these models are trained on non-medical images and hence do not preserve the complementary content of the input images resulting in the loss of information relevant to medical analysis.

In the context of the above mentioned drawbacks, this paper presents an end-to-end medical image fusion method using a content-aware GAN. It uses a GAN with one generator and two discriminators. The training phase involves subjecting the source image pairs to the generator which produces a fake fused image. The first and the second discriminator discriminates the fake image against source images one and two, respectively. The generator and discriminators are trained simultaneously against each other and as the training progresses the generator can generate better-quality fused images. During testing, the discriminators are discarded and the trained generator is utilized for fusing the source images. The paper is organized as follows. Section 7.2 gives the details of the proposed fusion method by highlighting the constituents blocks, network architecture, and loss functions used to train the GAN model. The experimental setting and results analysis is presented in section 7.3, and the conclusion is presented in section 7.4.

7.2 Proposed fusion method

The proposed end-to-end content-aware GAN-based multimodal image fusion method is shown in figure 7.1. The model is composed of two discriminators (D_1, D_2) and one generator. The generator architecture resembles the U-net having a series of downsampling and upsampling layers with skip-connections [24]. A total of eight downsampling units are used with one convolution and seven subsequent units comprising a convolution, batch-normalization, and LeakyReLU layer. This down-sampling block extracts the features from the source images and provides a sparse representation of their constituent information. Next, seven upsampling units are used, of which the first four consist of convolution, batch-normalization, dropout, and LeakyReLU layers, and the next three units consist of convolution, batch-normalization, and LeakyReLU layers. After each upsampling unit skip-connections are used for feature reuse from the previous downsampling units. In a generator model with k layers, the skip-connections are used between each layer j (starting from 0 to $k/2$) and layer $k-j$. Both D_1 and D_2 have the same architecture with two units each made of a convolution, dropout, and LeakyReLU layer followed by a flattened layer and a dense layer. The training flow is given in figure 7.1(a). Let A and B represent the pre-registered source CT and MR images, respectively, taken from the training dataset. These source images are concatenated along the channel dimension and given to the generator. The generator's goal is to generate images similar to both the source images and confuse the two discriminators D_1 and D_2. D_1 and D_2 aim to classify the generated fused image F as fake or real concerning images A and B, respectively. The conventional GANs use the min–max type of loss [20, 21, 24], however, in the case of fusion, to preserve the content that is the structures and texture of the input images, the loss function is redesigned as shown below.

The generator loss (L_G) consists of three parts namely intensity loss (L_{gan}), gradient loss (L_{grad}), and pixel loss (L_{pix}):

$$L_G = L_{gan} + L_{grad} + L_{pix}. \tag{7.1}$$

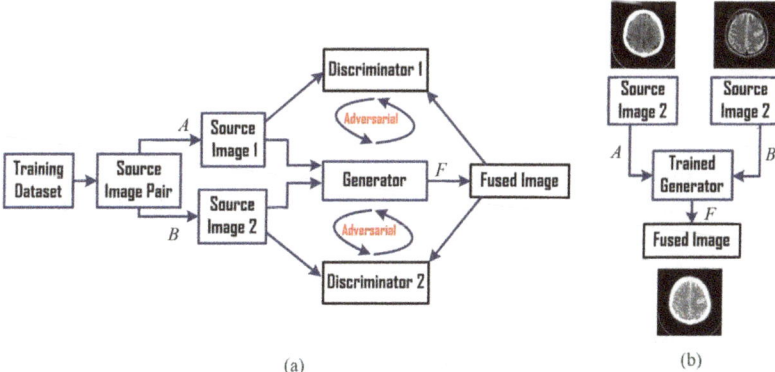

(a)   (b)

Figure 7.1. The proposed fusion method. (a) Training phase. (b) Testing phase.

L_G denoted the adversarial loss taking into account the losses of each of the discriminators [20, 21]:

$$L_{gan} = \mathbb{E}[\log(1 - D_A(F))] + \mathbb{E}[\log(1 - D_B(F))]. \tag{7.2}$$

The gradient loss aims at preserving the edge information of the tissue structures. As the MR image has better soft tissue demarcation and contains most of the edge information, the total variation norm consisting of the L_1 norm of the gradients along the x and y directions of the MR image is used to define the gradient loss as follows:

$$L_{grad} = \| F - B \|_{TV}. \tag{7.3}$$

The pixel loss aims to preserve the pixel intensities of the input images. As the pixel intensity relates directly to the type of tissue structures in both source images, it comprises the Frobenius norm of the maximum of the source images to retain the intensity details of pixels of both source images:

$$L_{pix} = \mathbb{E} \| F - \max(A, B) \|_{Fro}. \tag{7.4}$$

For training the two discriminators, the following optimization equations are used:

$$L_{D1} = \mathbb{E}[-\log(1 - D_1(A))] + \mathbb{E}[-\log(1 - D_1(F))] \tag{7.5}$$

$$L_{D2} = \mathbb{E}[-\log(1 - D_2(B))] + \mathbb{E}[-\log(1 - D_2(F))]. \tag{7.6}$$

During the testing as shown in figure 7.1(b), the source image pair are concatenated and subjected to the trained generator to obtain the fused image in an end-to-end manner. In the case of color image fusion, the functional images in RGB are converted to YUV color space to obtain one luminescence (Y) and two chrominance components (U and V) [11]. Next, the anatomical image and the Y component of the functional image are cascaded and given to the generator to obtain the fused Y component. Finally, the fused Y component and the original U and V components are cascaded and subjected to YUV to RGB color space conversion to obtain the final fused color image [11].

7.3 Experimental results and discussion

7.3.1 Dataset and implementation details

For training and testing the proposed model, a database from the Whole Brain Atlas of Harvard Medical School was used, consisting of neurological MR, SPECT, CT, and PET images [25]. For training, 242 pre-registered CT–MR image pairs are used with a spatial size of 256 × 256. The implementation was done in the TensorFlow framework, using the hardware platform with Intel(R) Core(TM) i7-4770, CPU, 3.40 GHz, 32-GB RAM, and 6GB NVIDIA GPU with 64-bit Windows operating system. The network is trained using Adam's optimizer for 20 epochs and the learning rate of 2×10^{-4}. The test dataset consists of 50 CT–MR, and 50 MR–SPECT/PET images belonging to various neurological disorders.

For validating the fusion performance, fused images are evaluated both visually and quantitatively. For quantitative performance analysis, nine state-of-the-art (SOTA) fusion metrics such as standard deviation (SD) [26], entropy (EN) [26], spatial frequency (SF) [26], edge preservation index ($Q_{AB/F}$) [27], Q_c, Q_y, mutual information (MI) [28], visual information fidelity for fusion (VIFF) [29], and the sum of the correlations of differences (SCD) [30] are used. To justify the efficacy of the proposed model the fusion results are compared with four existing DL based fusion methods, namely the dual-stream attention mechanism based method by Fu *et al* (Fu2021) [21], the dual-discriminator conditional GANs based method by Ma *et al* (Ma2020) [20], a dense net based unified fusion framework by Xu *et al* (Xu2022) [23], and a squeeze-and-decomposition network based method by Zhang *et al* (Zhang2021) [22].

7.3.2 Performance analysis

To demonstrate the efficacy of the proposed fusion method to generate good-quality fused images, a detailed visual comparison is presented. For this, three pairs of images are considered, as shown in figures 7.2(a) and (b). Further, the fused images obtained by four SOTA DL based fusion methods and the proposed method are also shown in figures 7.2(c) and (g). It is apparent from figures 7.2(a), (b), and (g) that for both gray and color image fusion cases, the proposed method fuses the complementary information of the source images effectively. For the CT–MR-T2 image fusion case, most of the methods fail to preserve the hard and soft tissue contrast and also result in visual artifacts, as shown in figures 7.2(c)–(f). However, the proposed method retains the soft tissue details and highlights the skull and other hard tissues as well. For color image fusion, it can be seen from figure 7.2 that the Ma2020 method loses the anatomical structures of the MR-T2 images completely. Although the Fu2021, Zhang2021, and Xu2022 methods preserve the edges of the soft tissues, the pixel intensity in the region of anomaly is not retained, resulting in color inconsistencies, as shown in figures 7.2(d)–(f) for the MR-T2–SPECT case. Also, for

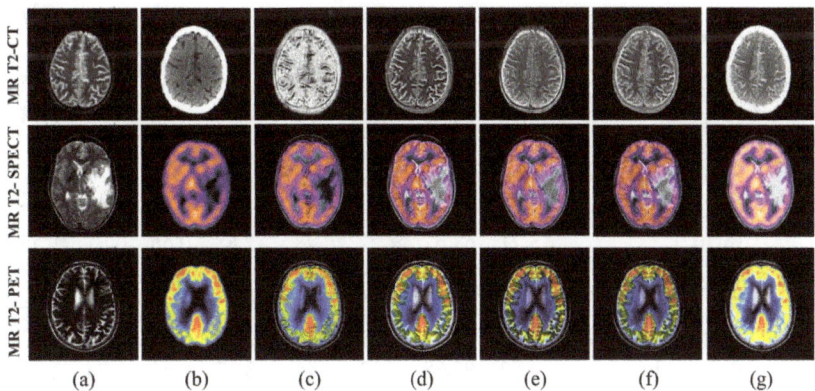

Figure 7.2. Subjective comparison of fusion results: (a) source image 1, (b) source image 2, (c) Ma2020, (d) Fu2021, (e) Zhang2021, (f) Xu2022, and (g) proposed fusion method.

MR-T2–PET image fusion these methods suffer from decreased contrast and cease to preserve the color distribution of the PET image. However, the proposed method works better in terms of the preservation of both the anatomical structures and functional color maps of the source images which can be visualized in figure 7.2 (g).

The quantitative results of the image pairs are shown in figure 7.2 and are also evaluated and presented in table 7.1.

The highest performing method is marked in bold and the second-highest is italicized. It can be seen from the results that the proposed method achieves higher values for most of the metrics, compared to the rest of the methods. Further, to demonstrate a more comprehensive performance comparison, the average values of all the metrics are calculated over the entire test dataset and listed in table 7.2. A rank is also assigned, with the highest performing method as rank 1 and the lowest performing method as ranked 5. From table 7.2 it is observed that the proposed fusion method ranks first for the metrics SD, MI, Q_Y, SCD, and VIFF. The Zhang2021 method ranks first for the $Q_{AB/F}$ metric which is a slight improvement, however, the proposed method outperforms the others for the rest of the metrics with a notable improvement in overall performance. The following is a performance summary base on table 7.2:

Table 7.1. Quantitative performance analysis of images shown in figure 7.2.

Image pair	Method	EN	SD	SF	$Q_{AB/F}$	MI	Q_C	Q_Y	SCD	VIFF
MR-T2–CT	Ma2020	*4.63*	89.52	**7.92**	0.25	2.71	0.53	0.39	*1.4*	0.27
	Fu2021	**5.02**	57.81	*7.74*	0.38	2.92	**0.67**	0.62	1.09	0.22
	Zhang2021	4.38	62.15	6.84	0.38	*3.07*	0.65	0.62	1.36	0.27
	Xu2022	4.31	58.5	6.86	**0.46**	2.98	0.64	*0.63*	1.18	*0.35*
	Proposed fusion method	4.14	**84.68**	6.00	*0.39*	**3.30**	*0.65*	**0.66**	1.53	**0.46**
MR-T2–SPECT	Ma2020	*5.06*	60.7	6.69	0.38	2.94	0.52	0.5	1.02	0.28
	Fu2021	**5.37**	*70.79*	**7.35**	*0.52*	3.06	*0.68*	0.67	1.51	0.5
	Zhang2021	4.48	68.34	*6.89*	**0.53**	3.09	**0.74**	**0.77**	*1.71*	0.54
	Xu2022	4.78	64.27	6.75	**0.53**	*3.21*	0.65	0.67	1.45	**0.57**
	Proposed fusion method	4.82	**84.40**	6.86	0.51	**3.40**	0.65	*0.70*	**1.78**	**0.65**
MR-T2–PET	Ma2020	*4.82*	63.16	7.1	0.35	2.85	0.56	0.57	1.24	0.29
	Fu2021	**5.21**	*63.67*	*7.77*	*0.48*	2.88	*0.66*	0.61	1.55	0.41
	Zhang2021	3.9	58.44	**7.78**	**0.52**	*3.03*	**0.74**	*0.73*	*1.65*	0.41
	Xu2022	4.22	53.72	6.62	0.42	3.01	0.61	0.57	1.47	*0.42*
	Proposed fusion method	4.53	**83.28**	7.62	0.44	**3.17**	0.65	**0.74**	1.73	**0.54**

Table 7.2. Averaged performance analysis (mean ± standard deviation (rank)).

Quantitative metric	MR image	CT/SPECT/ PET image	Ma2020	Fu2021	Zhang2021	Xu2022	Proposed method
EN	4.49 ± 0.65	3.8 ± 0.85	5.4 ± 0.7 (2)	5.41 ± 0.39 (1)	4.68 ± 0.71 (5)	4.74 ± 0.58 (4)	4.90 ± 0.59 (3)
SD	57.63 ± 7.4	71.49 ±16.32	74.14 ±16.71(2)	60.95 ± 5.48 (4)	61.7 ± 6.48 (3)	56.87 ± 7.12 (5)	81.11 ± 5.98(1)
SF	7.23 ± 0.74	5.2 ± 0.9	7.81 ± 0.73 (2)	7.9 ± 0.54 (1)	7.45 ± 0.71 (3)	6.97 ± 0.59 (5)	7.11 ± 0.68 (4)
$Q_{AB/F}$	—	—	0.37 ± 0.08 (5)	0.45 ± 0.09 (4)	0.5 ± 0.09 (1)	0.47 ± 0.05 (2)	0.46 ± 0.06 (3)
MI	—	—	2.73 ± 0.25 (5)	2.85 ± 0.27 (4)	3.05 ± 0.3 (2)	2.97 ± 0.29 (3)	3.25 ± 0.25 (1)
Q_C	—	—	0.55 ± 0.04 (5)	0.66 ± 0.05 (2)	0.69 ± 0.08 (1)	0.59 ± 0.06 (4)	0.65 ± 0.04 (3)
Q_Y	—	—	0.49 ± 0.07 (5)	0.63 ± 0.06 (3)	0.69 ± 0.1 (2)	0.59 ± 0.06 (4)	0.70 ± 0.04 (1)
SCD	—	—	1.27 ± 0.32 (5)	1.3 ± 0.27 (3)	1.48 ± 0.21 (2)	1.27 ± 0.25 (4)	1.60 ± 0.12 (1)
VIFF	—	—	0.27 ± 0.05 (5)	0.33 ± 0.17 (4)	0.37 ± 0.14 (3)	0.39 ± 0.11 (2)	0.48 ± 0.08 (1)
Average rank	—	—	4	2.8	2.4	3.6	2

1. The proposed method achieves 9.13% and 28.97% higher values of EN in the source images, signifying higher information content and making it more a appropriate and factual method for further analysis.
2. The proposed method gives 9.40%–42.62% higher values of SD than the source and fused images obtained from the other methods, indicating that it results in fused images with better contrast and tissue demarcation.
3. The proposed method demonstrates a higher ability to extract and integrate the crucial diagnostic information from the source images and obtains 6.55%–19.04% higher MI values than the other methods.
4. The proposed method also obtains 1.44%–42.85% and 8.10% and 25.98% higher Q_Y and SCD values than the other methods, reflecting its ability to generate fused images with higher similarity and correlation with source images compared to other methods.
5. The proposed method also achieves 23.07%–77.77% higher VIFF values than the other methods demonstrating that it performs better than the other methods in generating fused images with higher visual quality and less or no distortion.

Thus, from the performance analysis it can be concluded that the proposed fusion method provides effective fusion of multimodal medical images for both the gray and color image fusion cases. It also outperforms the existing DL based image fusion methods in both visual and quantitative performance with a notable improvement.

7.4 Conclusion

The proposed end-to-end content-aware GAN-based multimodal medical image fusion approach has a notably effective performance in the fusion of both gray and color image pairs. The loss function effectively characterizes the most meaningful information required to preserve the complementary characteristics of the multi-modal input images. The adversarial learning among the generator and two discriminators results in the effective synthesis of fused images with less information loss and improved visual quality in an end-to-end manner. Substantial experiments and analysis also justify the primacy of the proposed fusion method in highlighting a variety of tissue structures, and at the same time preserving the color details corresponding to the level of their metabolic activity, compared to the existing DL based fusion approaches. Detailed visual and quantitative performance analyses also validate that the proposed method encompasses a higher ability to fuse and represent various malignancies and anomalies compared to the source images and hence can assist radiologists in more accurate diagnosis and treatment.

References

[1] Hermessi H, Mourali O and Zagrouba E 2021 Multimodal medical image fusion review: theoretical background and recent advances *Signal Process* **183** 108036
[2] James A P and Dasarathy B V 2014 Medical image fusion: a survey of the state of the art *Inf. Fusion* **19** 4–19

[3] Dasarathy B V 2012 Information fusion in the realm of medical applications—a bibliographic glimpse at its growing appeal *Inf. Fusion* **13** 1–9

[4] Dogra A, Goyal B and Agrawal S 2017 Current and future orientation of anatomical and functional imaging modality fusion *Biomed. Pharmacol. J.* **10** 1661–3

[5] Du J, Li W, Lu K and Xiao B 2016 An overview of multi-modal medical image fusion *Neurocomputing* **215** 3–20

[6] Li S, Kang X, Fang L, Hu J and Yin H 2017 Pixel-level image fusion: a survey of the state of the art *Inf. Fusion* **33** 100–12

[7] Dogra A, Goyal B and Agrawal S 2017 From multi-scale decomposition to non-multi-scale decomposition methods: a comprehensive survey of image fusion techniques and its applications *IEEE Access* **5** 1640–67

[8] Meher B, Agrawal S, Panda R and Abraham A 2019 A survey on region-based image fusion methods *Inf. Fusion* **48** 119–32

[9] Wang Z, Wang S, Zhu Y and Ma Y 2016 Review of image fusion based on pulse-coupled neural network *Arch. Comput. Meth. Eng.* **23** 659–71

[10] Liu Y, Chen X, Liu A, Ward R K and Wang Z J 2021 Recent advances in sparse representation based medical image fusion *IEEE Instrum. Meas. Mag.* **24** 45–53

[11] Yang Y, Que Y, Huang S and Lin P 2016 Multimodal sensor medical image fusion based on type-2 fuzzy logic in NSCT domain *IEEE Sens. J.* **16** 3735–45

[12] Bashir R, Junejo R, Qadri N N, Fleury M and Qadri M Y 2019 SWT and PCA image fusion methods for multi-modal imagery *Multimedia Tools Appl.* **78** 1235–63

[13] Duan J, Mao S, Jin J, Zhou Z, Chen L and Chen C L P 2021 A novel GA-based optimized approach for regional multimodal medical image fusion with superpixel segmentation *IEEE Access* **9** 96 353–66

[14] Zhang H, Xu H, Tian X, Jiang J and Ma J 2021 Image fusion meets deep learning: a survey and perspective *Inf. Fusion* **76** 323–36

[15] Liu Y, Chen X, Wang Z, Wang Z J, Ward R K and Wang X 2018 Deep learning for pixel-level image fusion: recent advances and prospects *Inf. Fusion* **42** 158–73

[16] Hou R, Zhou D, Nie R, Liu D and Ruan X 2019 Brain CT and MRI medical image fusion using convolutional neural networks and a dual-channel spiking cortical model *Med. Biol. Eng. Comput.* **57** 887–900

[17] Liu Y, Chen X, Cheng J and Peng H 2017 A medical image fusion method based on convolutional neural networks *20th IEEE Int. Conf. on Information Fusion* pp 1–7

[18] Lahoud F and Süsstrunk S 2019 Zero-learning fast medical image fusion *22nd IEEE Int. Conf. on Information Fusion* pp 1–8

[19] Kazeminia S, Baur C, Kuijper A, van Ginneken B, Navab N, Al-bark S and Mukhopadhyay A 2020 GANs for medical image analysis *Artif. Intell. Med.* **109** 101938

[20] Ma J, Xu H, Jiang J, Mei X and Zhang X-P 2020 DDcGAN: a dual-discriminator conditional generative adversarial network for multi-resolution image fusion *IEEE Trans. Image Process.* **29** 4980–95

[21] Fu J, Li W, Du J and Xu L 2021 DSAGAN: a generative adversarial network based on dual-stream attention mechanism for anatomical and functional image fusion *Inf. Sci.* **576** 484–506

[22] Zhang H and Ma J 2021 SDNet: a versatile squeeze-and-decomposition network for real-time image fusion *Int. J. Comput. Vision* **129** 2761–85

[23] Xu H, Ma J, Jiang J, Guo X and Ling H 2022 U2Fusion: a unified unsupervised image fusion network *IEEE Trans. Pattern Anal. Mach. Intell.* **44** 502–18

[24] Isola P, Zhu J-Y, Zhou T and Efros A A 2017 Image-to-image translation with conditional adversarial networks *Proc. of the IEEE Conf. on Computer Vision and Pattern Recognition* pp 1125–34

[25] Summers D 2003 Harvard Whole Brain Atlas: www.med.harvard.edu/manila/home.HTML *J. Neurol. Neurosurg. Psychiatry* **74** 288–8

[26] Das M, Gupta D, Radeva P and Bakde A M 2020 NSST domain CT–MR neurological image fusion using optimized biologically inspired neural network *IET Image Proc.* **14** 4291–305

[27] Xydeas C and Petrovic V 2000 Objective image fusion performance measure *Electron. Lett.* **36** 308–9

[28] Qu G, Zhang D and Yan P 2002 Information measure for the performance of image fusion *Electron. Lett.* **38** 313–5

[29] Han Y, Cai Y, Cao Y and Xu X 2013 A new image fusion performance metric based on visual information fidelity *Inf. Fusion* **14** 127–35

[30] Aslantas V and Bendes E 2015 A new image quality metric for image fusion: the sum of the correlations of differences *AEU-Int. J. Electron. Commun.* **69** 1890–6

IOP Publishing

Data Analytics for Intelligent Systems
Techniques and solutions
Sachin Taran, Chhavi Dhiman and Manjeet Kumar

Chapter 8

Infrared thermography in diagnosing macular edema

J Persiya, A Sasithradevi and S Mohamed Mansoor Roomi

Infrared thermography helps to provide the temperature pattern of any region of interest in the form of thermal images captured by thermal cameras. These are called thermograms, and they make it possible to see the heat produced, which we cannot see with the naked eye. Some physiological changes in the human body can be monitored using thermal imaging. Hence thermal images are used widely in many medical applications. The correlation between the temperature and the pattern of blood flow makes it useful to diagnose conditions such as breast cancer, peripheral vascular disease, nerve damage due to diabetics, and fever. Since thermal imaging is a non-contact imaging method, it is a complementary method for ophthalmologists for the initial diagnosis and treatment of diseases. This study examines the diagnostic utility of infrared eye images for macular edema. This disease is the main cause of blindness and visual impairment worldwide. Infrared images of the retina using an infrared imaging camera are obtained to evaluate macular edema. This chapter discusses the importance of thermal image features in the diagnosis of macular edema. Different proposed methodologies for extracting thermal image features to classify healthy and non-healthy eyes are discussed. We conclude by highlighting the performance of machine and deep learning algorithms in diagnosing eye disease through thermal images and the future scope of such approaches.

8.1 Introduction

Macular edema (ME) is swelling in a part of the retina called the macula due to the leakage of blood vessels. The macula is a tissue layer at the back of the eye that is sensitive to light. ME is painless and usually at the initial stage it does not have symptoms. When the symptoms start, there will be signs that the blood vessels in the eye are leaking causing blurry vision. Diabetes is frequently linked to ME [4]. Figure 8.1 shows images of an unaffected eye and one with ME. ME is one of the

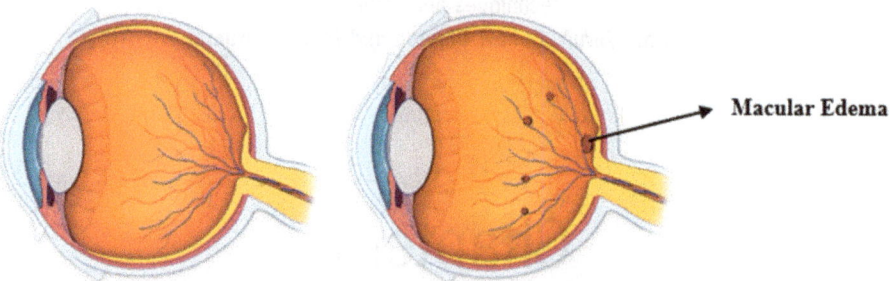

Figure 8.1. A normal eye and an eye affected by macular edema.

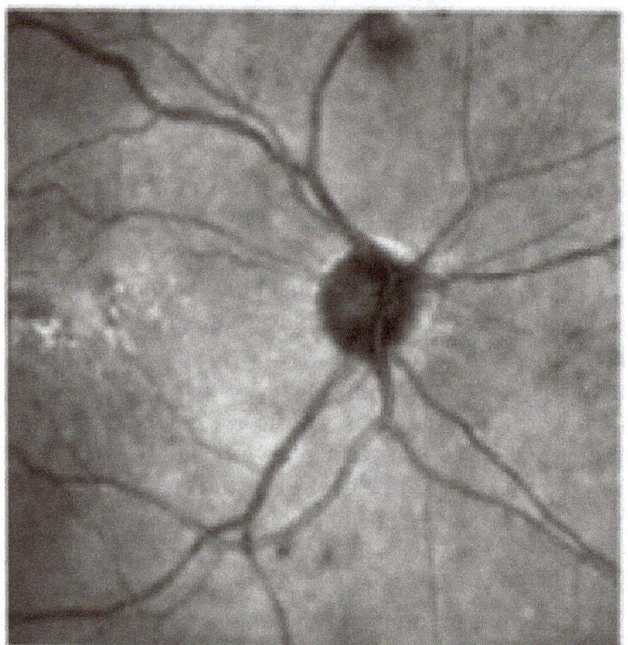

Figure 8.2. An IR image of an macular edema affected eye.

major causes of vision impairment in the world, mostly among patients affected by diabetes. Also, ME causes irreversible effects on eyesight. Hence adequate proper diagnosis and treatment of ME can reduce the risk of vision impairment [2].

Different eye imaging techniques are available. Techniques such as color fundus imaging, ocular coherence tomography (OCT), and fluorescein angiograms are used in the clinical diagnosis of macular edema [3]. Infrared (IR) thermography is gaining importance due to recent developments in infrared thermography in the realm of ophthalmology. Also, IR imaging is now performed as a regular step before OCT. An IR image of an eye with macular edema is depicted in figure 8.2. The main merit of using IR thermography is that it does not require a flash of light, as required in color fundus imaging, to take an image of the retina. Also, the fundus of the eye ball has the greater reflection of IR and IR can penetrate the sub-retinal layers and produce clear vessel contrast for detection of pathologies [26]. To take IR images, IR thermal cameras are required. The sensors in these cameras can detect infrared energy (radiated heat) to create visual images.

This chapter is organized into seven sections. Section 8.2 provides the physical principles of infrared thermal imaging. Section 8.3 introduces thermal imaging cameras. Section 8.4 explains the application of thermal infrared imaging in medicine. The procedures to be followed while capturing infrared images are explained in section 8.5. Section 8.6 explains briefly the different classification algorithms employed for thermal image classification between normal and diseased eyes. A detailed discussion is provided in section 8.7, while a conclusion is provided in section 8.8.

8.2 Physical principles of infrared thermal imaging

The thermal imaging technique is based on thermal physics. Thermal physics is the study of the analytical nature of any physical system from the perspective of energy. James Clerk Maxwell was a Scottish mathematician who was the first to demonstrate electromagnetic radiation. Electromagnetic waves are categorized by their frequency and wavelength, which are related to the speed of light. The following equation provides the radiation's frequency, where freq is the frequency, En is the energy and h is Planck's constant:

$$\text{freq} = \frac{\text{En}}{h}. \tag{8.1}$$

The electromagnetic spectrum depicts the typical distribution of electromagnetic waves absorbed or emitted by a specific object in relation to wavelength, frequency, and photon energy. Figure 8.3 shows the electromagnetic spectrum which ranges from low to high frequencies, including radio waves, microwaves, infrared radiation (IR), visible light, ultraviolet light, x-rays, and gamma rays.

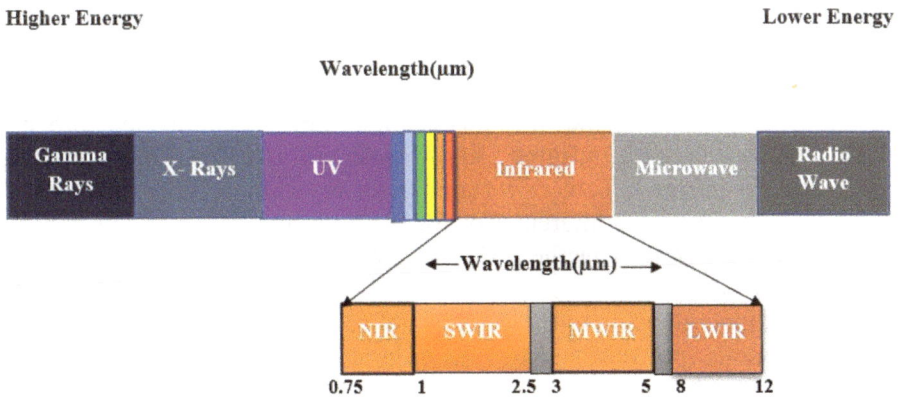

Figure 8.3. Electromagnetic spectrum.

Table 8.1. Regions within the infrared.

Various regions	Abbreviation	Wavelength	Frequency	Uses
Near-infrared	NIR	0.75–1.5 μm	215–400 THz	In estimating physical and anatomical constituents in manufacturing production.
Short-wavelength infrared	SWIR	1–2.5 μm	100–214 THz	In active illumination night vision technology.
Mid-wavelength infrared	MWIR	3–5 μm	37–100 THz	In cooled IR thermal cameras.
Long-wavelength infrared	LWIR	8–15 μm	20–37 THz	In uncooled IR thermal cameras.
Far-infrared	FIR	15 μm–1 mm	300 GHz–20 THz	In astronomy and therapeutic modalities.

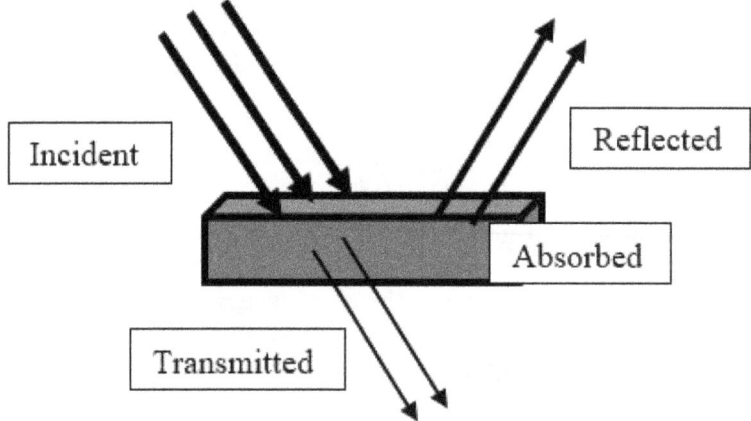

Figure 8.4. Effects of radiation.

As shown in figure 8.3, the infrared region is further sub-divided into near-infrared, short-wavelength infrared, mid-wavelength infrared, long-wavelength infrared, and far-infrared. Table 8.1 summarizes the abbreviations, wavelengths, frequencies, and uses of the different types of IR. Reflected infrared is the collective term for NIR and SWIR. Together, MWIR and LWIR are referred to as thermal infrared.

Thermal technology refers to the detection of radiation in the infrared region of the electromagnetic spectrum and the image produced by such energy. A type of infrared imaging is called thermography. As shown in figure 8.4 an object either absorbs, transmits, or reflects the incident radiation.

Hence

$$a + \imath + t = 1. \tag{8.2}$$

Here:

a is the coefficient of absorption $0 < a < 1$;
$\imath$ is the coefficient of reflection $0 < \imath < 1$;
t is the coefficient of transmission $0 < t < 1$.

To understand the relationship between temperature, emittance, and wavelength, according to Planck's rule, an object's spectral radiant emittance increases with temperature and decreases peak wavelength across the whole spectrum, which is given in

$$V_\lambda = \frac{C1}{\lambda^5[(e^{\frac{C2}{\lambda T}} - 1)]}, \tag{8.3}$$

where:

V_λ gives the spectral radiant emittance (W m^{-2} μm^{-1});
λ means the wavelength (μm) of the radiation;
T is the temperature of the blackbody (K).

The constant $C1 = 37\,418$ (W μm^4 m^{-2}) and the constant $C2 = 14\,388$ (μm K).
    The total radiation of the blackbody is derived by integrating Planck's law. The resulting equation is called the Stefan–Boltzmann law:

$$R = T^4\sigma, \tag{8.4}$$

where:

R is the radiant emittance (W m^{-2});
T is the temperature of the blackbody (K);
σ is Stefan–Boltzmann constant whose value is 5.67×10^{-8} (W m^{-2} K^{-4}).

According to the Stefan–Boltzmann law, a blackbody's radiant emittance varies as a function of its temperature to the fourth power.
    According to the Wien displacement law, differentiating Planck's law and equating it to zero determines the wavelength at which the spectral radiant exitance is highest:

$$\lambda_{\max} T = 2897.8 \,(\mu\text{m K}). \tag{8.5}$$

Not all sources of radiation are black bodies. Some of the energy that hits them may be sent, transmitted or reflected. So, a parameter called 'emissivity' can be used to describe any object. It is calculated as the ratio between the radiant emittance E^1 and the radiant emittance E of a blackbody at a certain temperature

$$\varepsilon = \frac{E^1}{E}. \tag{8.6}$$

In practice, emissivity values lie between 0 and 1. The blackbody radiation law states that all objects emit infrared radiation based on their temperature. Thermal imaging cameras calculate the temperature of an object over the operational wavelength range by detecting the emitted radiation. Emissivity influences the thermal image temperature calculation.

8.3 Infrared thermal imaging camera

The infrared spectrum cannot be seen by the human eye, so infrared cameras are required to take infrared images of it. These infrared cameras capture photos of the radiation they detect [9]. As an object's temperature rises, it emits more radiation. Thermography enables the observation of temperature fluctuations. Hence thermal cameras can be used to measure temperature in a non-contact manner.

ISO [16], IEC [15], and Howell and Smith [14] defined the minimum specifications for infrared cameras in medical applications. Proper protocols to use infrared cameras for different disease diagnoses must be followed. The most popular type of thermal detector that can be utilized when building thermal cameras is the focal plane array (FPA) built on thin-film bolometers. Thermal detector arrays are made to reduce the amount of heat lost by thermal conduction, enhancing the difference in temperature across the detector and hence its sensitivity. Infrared thermography's potential for use in medicine has been enhanced greatly by the creation of an uncooled, digital infrared camera [17]. There are two types of thermal imaging: passive and active. Short-wavelength infrared light is used by active or dynamic IR imaging systems to illuminate a target area. An image is created by the camera's interpretation of some of the infrared radiation that is returned to it. Systems for thermal imaging use IR light with a mid- or long wavelength. Devices called thermal imagers are passive and can only detect variations in heat. Images s a result of these heat disparities are then shown on a monitor. Since thermal imagers work at longer or mid-wavelengths than active IR, they are less influenced by their surroundings because they are unable to perceive reflected light.

8.4 Infrared thermal imaging in medicine

Imaging modalities such as x-rays, magnetic resonance imaging (MRI), and computed tomography (CT) have a good reputation for clinical utilization and reliability. However, when used repeatedly they cause a degree of health hazard due to their ionizing radiation, particularly for small babies and children. However, infrared radiation is non-ionizing radiation and therefore it will not cause cancer. The advancements in IR cameras along with IR detectors and image processing techniques have had a great influence on the field of medicine. Infrared thermal imaging is widely applied in examining burns, breast cancer, varicose veins, peripheral vascular disease, rheumatoid arthritis, etc.

In ophthalmology IR thermal images are used to detect diabetic eyes, dry eyes, glaucoma, thyroid eye disease, macular edema, and age-related macular degeneration (ARMD). Diseases such as ARMD and macular edema are diagnosed using IR images of the retina taken by active IR cameras [27], whereas diseases that can be detected over the ocular surfaces such as diabetic retinopathy, dry eye, glaucoma, etc, can be detected using passive thermal cameras. Regardless of changes in the surrounding environment, homeostatic processes attempt to maintain the body temperature. Arterioles in the dermis area allow more or less blood flow according to the exposed temperature [17]. As temperature is correlated with the blood flow pattern over the surface of the body, thermal cameras can produce the temperature pattern of any region of interest. Hence it can be used in the diagnosis and monitoring of various diseases.

8.5 Procedure for infrared imaging

Mapstone [18] examined the variables influencing corneal temperature and found them to be the outside temperature, blinking [20], the carotid artery [19], anterior uveitis, aqueous humor, and uneven blood flow to the anterior portions [17]. The major prerequisite for using IR imaging in clinical settings is to keep the room at a temperature between 18 °C and 25 °C [23, 24]. Participants are asked to arrive for IR imaging without any topical cosmetics or ointments. Smoking and drinking must both be avoided. It is best to refrain from drinking hot tea and coffee [5]. Patients are requested to come and rest for 10–15 min to achieve adequate stability in blood pressure and skin temperature [6]. Patients are supposed to sit casually without bare feet. Infrared cameras must be switched on 15 min before imaging and properly mounted on the camera stand. The details of the participants should be recorded with the IR images to avoid misinterpretations. A proper procedural IR imaging technique improves the accuracy of diagnosis of eye diseases.

8.6 Classification algorithms used for thermal imaging

Advancements in the use of algorithms for machine learning and deep learning paved the way for their use in various automated medical diagnosis methods. Various classification algorithms are used in various medical infrared thermal images. Different algorithms depend on different types of input variables. The region of interest in the photographs serves as the source of the input variables, which can be either first-order statistics or second-order statistics. Following that, the classifier is assessed according to its accuracy, specificity, and sensitivity. Some of the commonly used effective classification algorithms are explained briefly below.

8.6.1 Artificial neural network (ANN)

One input layer, one output layer, and two or more hidden layers make up an artificial neural network (ANN). Each layer is connected, and each connection is associated with a weight. By adjusting the weight, the neural network learns. The network performance improves by updating the weight iteratively. Since ANN learns the training data, the error is reduced. Figure 8.5 shows the ANN structure which consists of neurons and interconnections.

The connection between the nodes has weights. Each node's input values have their associated weights applied to scale them. IN ANN the function is sent from the input nodes to the output nodes. Iteratively throughout the learning process, the

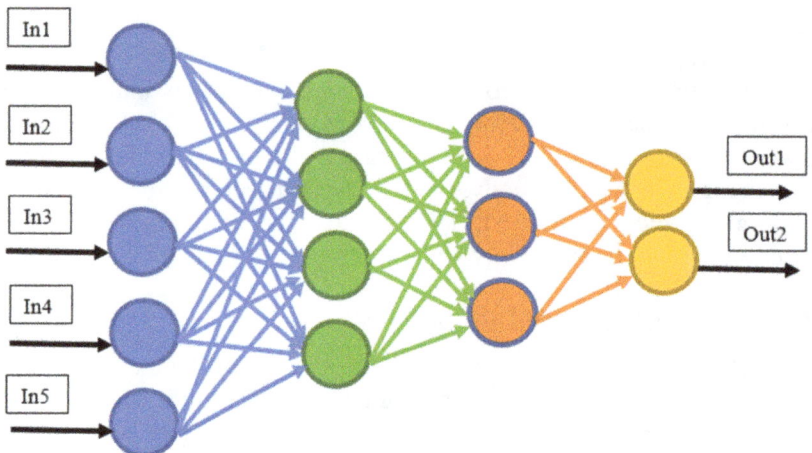

Figure 8.5. ANN.

weights, which are intermediate parameters, are adjusted [13]. The output from one node in a simple ANN network is sent to another node. The messages are transmitted in this manner across the layers of nodes that connect one another. More complex behaviors might be modeled with feedback that has a high non-linearity. The neural network is somewhat similar to the network of neurons in the human brain. In order to mimic human intelligence, an ANN combines the objectives of machine learning techniques with the complexity of some statistical methodologies.

8.6.2 Support vector machine (SVM)

The most used classifier for medical thermal images is the SVM. It can be used for both linear and non-linear problems. Figure 8.6 shows the SVM classifier. Binary linear outcomes are produced by SVM classifiers. One or two potential categories are assigned when the SVM classifier analyses the initial data. SVM uses the mapping of points in space to represent the model. Then a hyperplane is created to separate the two classes. The points are distributed according to the class to which the points belong. Maximizing the margin is the primary optimization goal in SVMs. The margin is described as the separating area between the hyperplane, i.e. the decision boundary, and the data points lying on the separation, which are called support vectors. This illustrated in figure 8.6. SVM selects the most efficient hyperplane to classify the data. It is efficient for both small and large datasets.

8.6.3 Naive Bayes (NB)

The naive Bayes classifier is a probabilistic classifier that operates under the presumption of input source independence. An algorithm based on probabilities serves as the classification method for NB. It uses numerical data to operate and

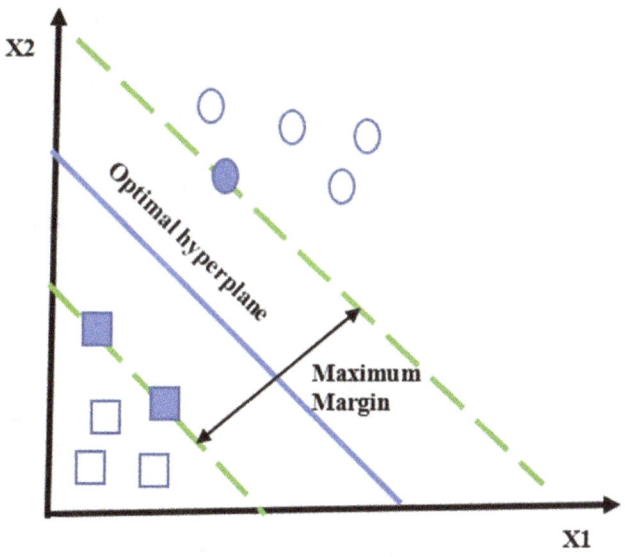

Figure 8.6. SVM.

when the categories are combined, the data produce a probabilistic model with a foundation on the application of Bayes' theorem. Bayes' theorem in mathematical terms is

$$P(c \mid d) = \frac{P(c) \cdot P(d \mid c)}{P(d)}, \tag{8.7}$$

where c and d are two different events, and $P(c)$ and $P(d)$ are the prior probabilities of c and d without reference to one another, i.e. they are independent. $P(c|d)$ is called posterior probability which denotes the probability of present event c given that d is true. $P(d|c)$, also known as likelihood, is the probability of present event d given that c is true.

Assuming that e_j denotes one of K classes and that vector $Y = (y_1, y_2, \ldots, y_n)$ is an instance with n independent attributes that need to be classified, we can use the Bayes theorem to determine the posterior probability, $P(e_j|Y)$, using $P(e_j)$, $P(Y)$, and $P(Y|e_j)$. The naive assumption made by the naive Bayesian classifier is known as class conditional independence, which states that the impact of a predictor's value (y_i) on a particular class (e_j) is independent of the values of other predictors. For the classification prediction, a limited amount of numerical data is needed. As a result, it is a quick and practical classification algorithm. A small amount of data is needed for both training and categorization. However, even though it is a fast algorithm, it shows low accuracy [1].

8.6.4 AdaBoost

Adaptive boosting is a standard method used in several models. Prior to making predictions on the training set, a simple or weak classifier is trained. The classifier is trained using the updated weight while the weight of the incorrectly classified set is increased. In order to generate a good classifier, it is made to predict once more on the training set. The steps followed in the concept of AdaBoost apply an initially uniform weight to the entire training set. During the boosting process, the weighted weak classifier is trained and the class labels are determined. Next, the coefficient and weighted error rate are calculated. The weights are updated and the weights' normalization is set to 1. Finally, the final prediction is computed. This classifier is sensitive to data and noise [10].

8.6.5 k-nearest neighbors (k-NN)

Among different machine learning algorithms, the k-NN approach is one of the easiest for classifying objects. 'k' implies the closest training sample near the classification model. It is useful to assign weights to the neighbors, hence the nearest neighbor contributes more to the average of the model compared to those far away. The best metrics must be chosen, i.e. how to calculate the distance between two nearby points, in order to choose this approach. There may not be a problem with some datasets, but multivariate data, where measurements are made on various scales, usually requires some standardization. This is typically interpreted as either the variable's standard deviation or its range. Metrics can occasionally also depend on various classes, allowing one to calculate distances conditionally on the class. While processing and classification times will lengthen, there may be a significant improvement in performance as a result. For classes with few samples, the regularized value is used in case a trade-off exists between the rescaling parameters of the within-class value and the global value. Cross-validation techniques, in which the training data are divided into two parts and the second portion is classified using a k-NN rule, can be used to make the decision. The method is time-consuming in large datasets, however, because each classification requires storing and looking over all of the training data. k-NN is a nonparametric model built on instance-based learning. The k-NN algorithm is an easy-to-understand algorithm. It begins by picking a distance measure and the number of k. The next step is to identify the sample's k-nearest neighbors that require classification. After all voting is completed, a class label will be assigned. Figure 8.7 shows how the moon class is labeled based on the majority of voting in seven nearest neighbors.

In the training dataset, the k-NN algorithm locates k samples that are close to the classification boundary and are based on the selected metrics. Next, the new data point's k closest neighbors vote to decide what class label to assign it, with the majority winning. Finding the ideal value for k is crucial for maintaining a reasonable balance between under- and overfitting:

$$E(x, y) = \sqrt[p]{\sqrt{\sum_{i=0}^{n}(x_i - y_i)^p}} .$$ (8.8)

For real-valued samples, the most common distance measure is the Manhattan distance and Euclidean distance measure, which is provided in equation 8.8. x and y

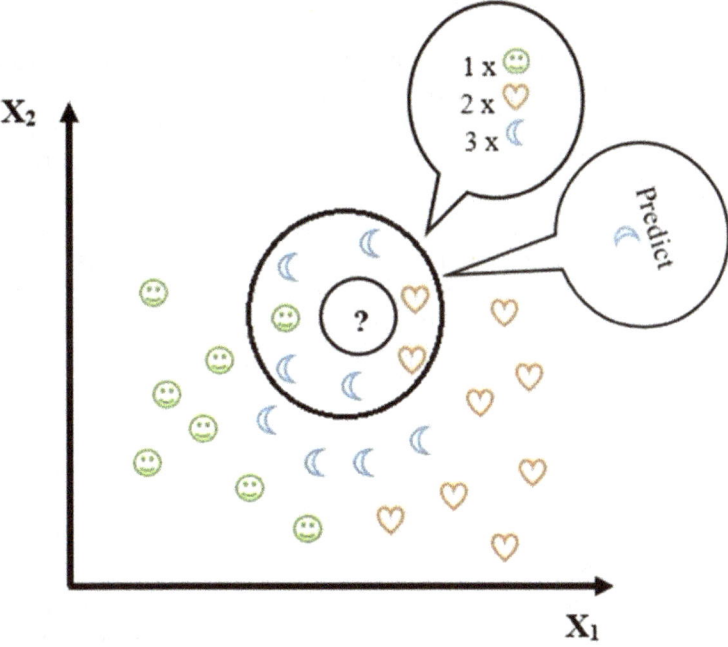

Figure 8.7. k-nearest neighbors.

are neighboring points of n samples. It is important to choose a distance measure that complements the dataset's attributes.

8.6.6 Decision tree (DT)

A supervised machine learning approach called the decision tree is utilized for both classification and regression. It classifies the data and shows the result in the form of a flow chart-like tree structure. It classifies the data based on some attributes called root nodes at the top and is further divided into leaf nodes that hold the class name. During the training process, based on the training data, a decision tree is constructed. The decision tree is designed to maximize the amount of information gained (IG). The main primary goal is to maximize the information acquired at each split; the function is

$$IG(D_R, f) = I(D_R) - \sum_{i=1}^{m} \frac{N_i}{N_R} I(D_i). \tag{8.9}$$

Here f is the feature attribute to perform the split, D_R and D_i are the datasets of the root and the ith child node or leaf node, I is the impurity measure, N_R is the overall number of samples at the root node, while N_i is the count of samples in the ith child node. The difference between the root node's impurity and the total of the impurities on the leaf nodes is the information gain. The information gain increases as the leaf nodes' impurity decreases. The main drawback of this method is that overfitting may occur during the construction of the decision tree.

8.6.7 Random forest

The most popular and effective algorithm of machine learning techniques is random forest. It consists of several DTs. A forest is created by developing a greater number of decision trees. Tree growth is made more random using the random forest algorithm. Different training examples are used to train each decision tree. The random forest determines the DT with a greater number of votes, thus yielding a better classifying model. In most cases, there is no need to prune the random forest because the ensemble model is extremely resilient to noise from the individual decision trees. The quantity of trees that must be selected for the random forest is a crucial element for the random forest method. Generally speaking, the random forest classifier performs better the more trees there are, albeit at the expense of a higher computing cost.

8.7 Discussion

The comparison of classifiers used for infrared thermal image diagnosis of various eye disorders is shown in table 8.2. The common methodology used for the diagnosis of eye diseases using a machine learning algorithm is depicted in figure 8.8.

The IR image acquisition is followed by image pre-processing. The crucial phase in the automatic diagnosis of illnesses is feature extraction. Statistical texture features are very much useful in the classification of images in the gray-level spatial distribution. The various approaches to statistical texture analysis include the first-order statistics, the second-order statistics, and the higher-order statistics. Only one

Table 8.2. Comparison of classifiers for different eye diseases.

Author name	Disease	Number of images	Method	Performance metrics used
Azharuddin et al [7]	Dry eye	42 diseased and 36 healthy	Statistical method	–
Selvathi et al [25]	Diabetic eye	149 diseased and 134 normal	SVM	Accuracy - 86.22%, Sensitivity - 91.79%, Specificity - 79.83%
Ajaz et al [2]	Macular edema	23 diseased and 18 normal	SVM, k-NN, naive Bayes	KNN & SVM Sensitivity - 100%, Accuracy - 100% Naive Bayes Sensitivity - 100%, Accuracy - 97.6%
Ujalambe et al [28]	Glaucoma	50 diseased and 50 normal	SVM	Accuracy - 95%
Padmapriya et al [22]	Glaucoma	15 normal and 15 diseased	SVM	–
Chandrasekar et al [8]	Diabetic eye	80 normal and 70 diseased	Statistical method	–

| Harshvardhan et al [12] | Glaucoma | 15 normal and 15 diseased | Logistic regression | Sensitivity - 60.6 %, Specificity - 70.3%, Accuracy - 88.8% |
| Acharya et al [1] | Dry eye | 83 diseased and 21 normal | k-NN, naive Bayes, DT, probabilistic neural network, and SVM | Left eye using PNN and KNN classifiers Accuracy - 99.8%, Sensitivity - 99.8%, Specificity - 99.8% Right eye using SVM classifier Accuracy - 99.8%, Sensitivity - 99.9%, Specificity - 99.4% |

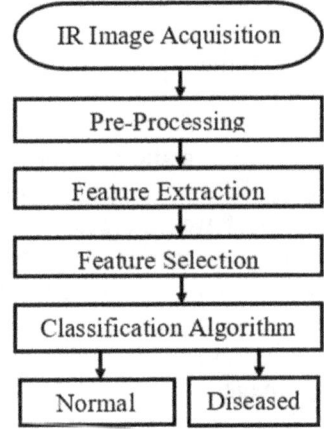

Figure 8.8. Methodology.

pixel is used in first-order statistics, two pixels in second-order statistics, and more pixels in higher-order statistics (HOS). HOS are also used in non-Gaussian and non-linear pictures. First-order statistical features including mean, skewness, entropy, kurtosis, variance, and energy are extracted using histogram texture features. Using gray level co-occurrence matrix (GLCM) based texture features is a method to extract second-order statistical features such as entropy, autocorrelation, correlation, energy, contrast, etc. Haralick features can be extracted using GLCM and are commonly used in many medical imaging approaches [11]. After feature extraction appropriate attributes must be selected to obtain high accuracy. Finally, classification algorithms are used to classify diseased images from normal images.

Selvathi *et al* [25] used GLCM based features based on RGB and HOS images extracted after the thermal images have been analysed, and an SVM classifier was utilized for categorization and obtained an accuracy of 86.22%. Ajaz *et al* [2] used histogram and GLCM texture characteristics and used different classifiers and found that SVM and k-NN were shown to have superior accuracy when compared to naive Bayes. Acharya *et al* [1] extracted the features from higher-order spectra, a non-linear technique. Then used different classifiers such as k-NN, naive Bayes, SVM, probabilistic neural network, and DT, and they found that k-NN and SVM showed a high accuracy of 99.8%. Harshvardhan *et al* [12] used the logistic regression method to classify normal images from glaucoma affected eye images and obtained an accuracy of 88.8%. Ujalambe *et al* [28] used GLCM for feature extraction of the gray images and used an SVM classifier and obtained an accuracy of 95%. In the comparison of the studies table 8.2, the SVM classifiers reveal greater accuracy in distinguishing between normal and diseased eyes. There is no publicly accessible dataset for the various eye conditions that different authors have used. All of the datasets that different authors have used to diagnose diseases are private.

8.8 Conclusion

Automated pathology diagnosis and screening are required due to the increase in patients with eye disorders. To prevent visual loss and impairments caused due to macular edema, early detection of the disease is required [21]. Long-term diabetes often results in blood vessel deterioration and blood leakage, which are not visible in fundus photographs. Therefore, IR retinal images aid ophthalmologists in accurately diagnosing ME. Since only fewer samples were used for this study, more samples would need to be added in order to evaluate the findings further. Deep learning algorithms are also being developed for the classification of medical images, and they demonstrate extremely high accuracy. Therefore, in the future, deep learning algorithms can be used to assess and enhance the accuracy of diagnosing macular edema using infrared images.

References

[1] Acharya U R, Tan J H, Koh J E, Sudarshan V K, Yeo S, Too C L, Chua C K, Ng E Y K and Tong L 2015 Automated diagnosis of dry eye using infrared thermography images *Infrared Phys. Technol.* **71** 263–71

[2] Ajaz A and Kumar D K 2020 Infrared retinal images for flashless detection of macular edema *Sci. Rep.* **10** 1–11

[3] Ajaz A, Kumar H and Kumar D 2021 A review of methods for automatic detection of macular edema *Biomed. Signal Process. Control* **69** 102858

[4] Aliahmad B, Kumar D K and Jain R 2016 Automatic analysis of retinal vascular parameters for detection of diabetes in Indian patients with no retinopathy sign *Int. Scholar. Res. Not.* **2016** 8423289

[5] Ammer K 1997 The influence of antirheumatic creams and ointments on the infrared emission of the skin Abs *10th Int. Conf. Thermogrammetry and Thermal Engineering 18–20 June 1997, Budapest* I Benkö *et al* (Budapest: MATE)) pp 177–81

[6] Ammer K and Ring E F J 2006 Standard procedures for infrared imaging in medicine *Biomedical Engineering Handbook* (Boca Raton, FL: CRC Press) p 1

[7] Azharuddin M, Bera S K, Datta H and Dasgupta A K 2014 Thermal fluctuation-based study of aqueous deficient dry eyes by non-invasive thermal imaging *Exp. Eye Res.* **120** 97–102

[8] Chandrasekar B, Rao A P, Murugesan M, Subramanian S, Sharath D, Manoharan U, Prodip B and Balasubramaniam V 2021 Ocular surface temperature measurement in diabetic retinopathy *Exp. Eye Res.* **211** 108749

[9] Elsner A E, Burns S A, Hughes G W and Webb R H 1992 Reflectometry with a scanning laser ophthalmoscope *Appl. Opt.* **31** 3697–710

[10] EtehadTavakol M, Chandran V, Ng E Y K and Kafieh R 2013 Breast cancer detection from thermal images using bispectral invariant features *Int. J. Therm. Sci.* **69** 21–36

[11] Haralick R M, Shanmugam K and Dinstein I H 1973 Textural features for image classification *IEEE Trans. Syst. Man Cybern* **SMC-3** 610–21

[12] Harshvardhan G, Venkateswaran N and Padmapriya N 2016 Assessment of glaucoma with ocular thermal images using GLCM techniques and logistic regression classifier *Int. Conf. on Wireless Communications, Signal Processing and Networking (WiSPNET)* (Piscataway, NJ: IEEE) pp 1534–7

[13] Hasan D A, Zeebaree S R, Sadeeq M A, Shukur H M, Zebari R R and Alkhayyat A H 2021 Machine learning-based diabetic retinopathy early detection and classification systems—a survey *1st Babylon Int. Conf. on Information Technology and Science (BICITS)* (Piscataway, NJ: IEEE) pp 16–21

[14] Howell K J and Smith R E 2009 Guidelines for specifying and testing a thermal camera for medical applications *Thermol. Int* **19** 5–14

[15] IEC 80601-2-59:2017 Medical electrical equipment—Part 2–59: particular requirements for the basic safety and essential performance of screening thermographs for human febrile temperature screening

[16] ISO/TR 13154:2017 Medical electrical equipment—deployment, implementation and operational guidelines for identifying febrile humans using a screening thermograph

[17] Jones B F 1998 A reappraisal of the use of infrared thermal image analysis in medicine *IEEE Trans. Med. Imaging* **17** 1019–27

[18] Mapstone R 1968 Determinants of corneal temperature *Br. J. Ophthalmol.* **52** 729

[19] Mapstone R 1968 Normal thermal patterns in cornea and periorbital skin *Br. J. Ophthalmol.* **52** 818

[20] McMonnics C W, Korb D R and Blackie C A 2012 The role of heat in rubbing and massage-related corneal deformation *Cont Lens Anterior Eye* **35** 148–54

[21] Medhi J P, S.R N, Choudhury S and Dandapat S 2023 Improved detection and analysis of macular edema using modified guided image filtering with modified level set spatial fuzzy clustering on optical coherence tomography images *Biomed. Signal Process. Control* **79** 104149

[22] Padmapriya N, Venkateswaran N, Kannan T and Madhuri M S 2015 Assessment of Glaucoma with ocular thermal images using GLCM techniques *12th Int. Conf. on Quantitative Infrared Thermography (Mamallapuram, India)*

[23] Ring E F J and Ammer K 2012 Infrared thermal imaging in medicine *Physiol. Meas.* **33** R33

[24] Ring E F J and Ammer K 2015 The technique of infrared imaging in medicine *Infrared Imaging: A Casebook in Clinical Medicine* (Bristol: IOP Publishing)

[25] Selvathi D and Suganya K 2019 Support vector machine based method for automatic detection of diabetic eye disease using thermal images *1st Int. Conf. on Innovations in Information and Communication Technology (ICIICT)* (Piscataway, NJ: IEEE) pp 1–6

[26] Schmitz-Valckenberg S, Steinberg J S, Fleckenstein M, Visvalingam S, Brinkmann C K and Holz F G 2010 Combined confocal scanning laser ophthalmoscopy and spectral-domain optical coherence tomography imaging of reticular drusen associated with age-related macular degeneration *Ophthalmology* **117** 1169–76

[27] Sidibe D *et al* 2017 An anomaly detection approach for the identification of DME patients using spectral domain optical coherence tomography images *Comput. Methods Programs Biomed.* **139** 109–17

[28] Ujalambe S J and Madhe S P 2021 Glaucoma detection using the thermal image processing *5th Int. Conf. on Intelligent Computing and Control Systems (ICICCS)* (Piscataway, NJ: IEEE) pp 761–5

IOP Publishing

Data Analytics for Intelligent Systems
Techniques and solutions
Sachin Taran, Chhavi Dhiman and Manjeet Kumar

Chapter 9

Variants of generative adversarial networks for underwater image enhancement

M Vijayalakshmi, A Sasithradevi and P Prakash

Underwater image enhancement is the most important technique used in numerous applications such as aquaculture, deep sea exploration, pipeline maintenance and inspection, structural defect analysis, and underwater topography. Owing to its importance, deep learning techniques are used widely for applications related to underwater image enhancement. Among the most common deep learning techniques, the generative adversarial network is commonly used for enhancement purposes due to its characteristics such as image-to-image translation, boosting the resolution of images, and generating synthetic and high-fidelity images. In this chapter we first introduce the formation model for optical images in an underwater environment to aid the understanding of the underwater image degradation process. Then the fundamental reasons for degradation, such as physical limitations of the camera and unsuitable lighting conditions, are discussed. Then various generative adversarial network (GAN) based underwater image enhancement models, such as the underwater generative adversarial network (UGAN), WaterGAN, underwater generative adversarial network (UWGAN), underwater image enhancement-stacked GAN (UIE-sGAN), multiscale cycle generative adversarial network (MCycleGAN), fusion generative adversarial network (FGAN), and DenseGAN are studied to acquire knowledge about the architecture, data training, training configuration, optimizers, and the loss function and its parameters. The evaluation metrics and underwater image datasets are also summarized. Finally, several open issues in underwater image enhancement and future research scope are discussed. We hope this chapter might provide a source for developing deep GAN-based underwater image enhancement and enhance the future scope of research work.

9.1 Introduction

In recent times the underwater environment has evolved into a great source of interest for researchers across the world as it contains valuable and abundant resources. The whole economy relies on these vast underwater resources. However, it's noteworthy that only a small percentage of these resources are utilized for underwater research purposes. To explore the underwater environment, underwater images (UWIs) and videos are used. Some high-resolution UWIs are shown in figure 9.1. Since UWIs suffer from huge challenges compared to generic images, it is very important to preserve the quality/clarity of images or video which will help in easy and effective information processing. UWIs are optical images as these images are captured using optical cameras. Underwater optical images face various challenges such as absorption [3], light illumination [2] scattering, geometric and photometric distortion [1], color degradation [4], noise distribution [5], and water turbidity, which in turn results in a hazy appearance [6]. These challenges pave the way for researchers to contribute novel creations to address these issues.

The design of a deep learning architecture mainly depends on the image formation models that cause degradation in the underwater images analysed. As underwater is a more complex environment, the image formation for the underwater model differs greatly from the atmospheric model [7]. The atmospheric image formation model mainly depends on three components, namely, the direct light reflected by the camera, a transmission map that represents the distance between the camera and the target, and the atmospheric light illumination. In the underwater model [7], the image formation depends on the light directly reflected from the target, called forward scattering, backward scattering from particles in the water, and light illumination from natural sources. In most models the backward scattering can be considered a negligible parameter. The major difference between atmospheric and underwater images formation is light illumination, i.e. in the atmosphere the natural light source provides uniform illumination, whereas underwater illumination depends on depth. As depth increases light becomes attenuated and higher wavelength light is absorbed, whereas shorter wavelength light travels deeper in water [8]. This is the reason why the sea has a bluish/greenish color. Figure 9.2 demonstrates image formation in an underwater environment.

Figure 9.1. Sample high-resolution UWI and corresponding degraded image from the EUVP dataset [43].

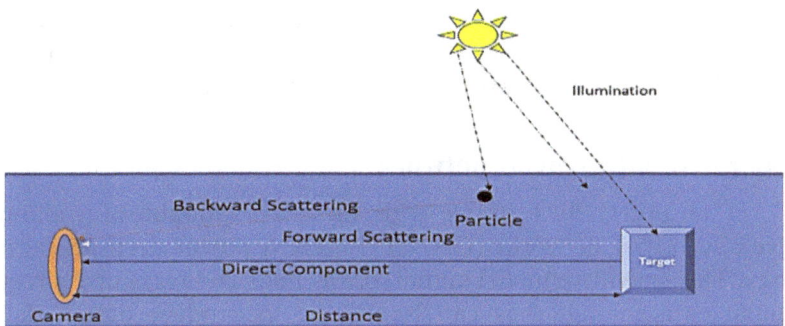

Figure 9.2. Underwater image modeling considering direct reflected light, and forward and backward scattering.

9.2 Background and key concepts

In this section the need for image restoration for an UWI with a degraded format is discussed. To understand the process of UWI degradation, various image formation models proposed by researchers [7] are analysed. To perform high-level image processing on underwater imagery [4] the first and foremost process is restoring degraded images [5]. The performance metrics can be relatively high for high-level tasks by accurately implementing low-level tasks.

9.2.1 The atmospheric model for image formation

The majority of UWI restoration is based on modified versions of the classic model for image creation, such as the atmospheric scattering model [7]. In an underwater environment, the light directed from the image acquisition system to the target experiences both absorption and scattering effects. The primary elements of the image captured depend on forward scattering from the target, backward scattering, and the light being reflected by additional water-based particles, thus turbidity also plays a vital role. Light attenuation is also a cause of degraded images, as shorter wavelengths can reach greater depths and longer wavelengths can travel only a few meters [8]. Thus the algorithm should be designed in such a way that all the above factors should be considered.

The classic atmospheric scattering model [9] is derived as

$$x = V(y)(P(y)) + I(1 - P(y)), \tag{9.1}$$

where $V(y)$ is the observed image taken in an underwater environment, y is the undistorted image, $P(y)$ refers to the exponential decay term with respect to attenuation, and I refers to the ambient light intensity level. $P(y)$ can be expressed as below if a uniform haze is observed:

$$P(x) = \exp(-\beta d(y)), \tag{9.2}$$

where $d(y)$ is the distance from the scene to the camera and β refers to the attenuation coefficient of light in the atmospheric environment. In this scattering model the light attenuation is taken as independent variable.

9.3 Generative adversarial network

To perform unsupervised learning with the least amount of training data, generative adversarial networks (GANs) [10] are a potent deep neural network technology. The generative model automatically finds and picks up on regularities in the input data so that they can be utilized to produce output data. This generative modeling is a machine learning appraoch that involves the GAN structure and can be split into three parts: generative, adversarial, and network. Learning the generative model that represents the process of data production in probabilistic terms is the primary purpose of the generative model. The adversarial part is used to train the model based on the adversarial setting. The deep neural network is used as the algorithm for training the entire network. GANs are an efficient way to train the generative model, which is composed of both the generator model and the discriminator model [11].

The generator generates fake sample data, such as audio, image, etc, and the generator also tries to falsify the discriminator [1]. The discriminative model tries to distinguish whether the image is real or fake. First, the random noise y is sampled using uniform or normal distribution. With y as the input the generator X creates an image I where $I = X(y)$). The generator model [13] creates only random noise and subsequently the discriminator [15] provides guidance to the generative model about how to create the image. The framework for the GAN is shown in figure 9.3.

While the discriminator attempts to maximize the loss function, the generator's primary goal is to decrease loss [12]:

$$\min_K \max_J I(J, K) = N_{y \sim p(y)}[\log L(y)] + N_{x \sim p(x)}[\log (1 - L(K(x)))], \qquad (9.3)$$

in which:

$L(y)$ is the probability estimation of the discriminator that real data instance y is real;

N_y are the real instances of data for the expected value;

$K(x)$ is the generator's output when given noise x;

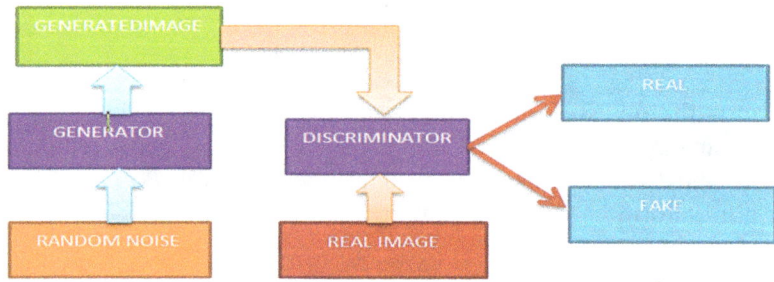

Figure 9.3. Framework for the GAN.

$L(K(x))$ is the probability estimation of the discriminator that the real data instance is fake;

N_x is the expected value over all random inputs to the generator.

The algorithm used in the minimax game is straightforward and is briefly explained in the section below. This demonstrates clearly the algorithm's attempt to use the back-propagation system to optimize the gradients.

Algorithm. for n number of iterations do

for k (hyperparameter)

Step 1.

- Sample n noise sample in minibatch $\{x^{(1)}, ..., x^{(a)}\}$ from noise prior $p(x)$.
- Sample n examples in minibatch $\{y^{(1)}, ..., y^{(a)}\}$ from the distribution $p(y)$.
- Update the discriminator using stochastic gradient descent in an ascending manner.

$$\nabla_{\theta 1} \frac{1}{n} \sum_{i=1}^{m} \log L(y^{(i)}) + \log Z(1 - L(P(x^{(i)})))].$$

end for

Step 2.

- Sample n noise sample in minibatch $\{x^{(1)}, ..., x^{(a)}\}$ from noise prior $p(x)$.
- Update the discriminator using stochastic gradient descent in an ascending manner.

$$\nabla_{\theta k} \frac{1}{n} \log Z(1 - L(P(x^{(i)}))).$$

end for

Update the gradients using any standard gradient learning rule.

9.4 Variants of GAN for image enhancement

Unsupervised learning is used in GANs, a model that uses two neural networks and has implicit density estimation. In this section, we explain briefly the various GAN architectures developed by various authors in the field of UWI enhancement.

9.4.1 DenseGAN

The multiscale dense block (MSDB) technique is used by DenseGAN [13] to improve underwater images that include color aliasing and underexposure. The generator block consists of a multiscale dense block in which the multiscale is used to increase the performance level, dense concatenation is used to collect more details from the original image, and residual learning is used for learning the existing features. The function of the discriminator block is to stabilize the learning by incorporating light computation [4] and spectral normalization. Combining the $L1$

loss and gradient loss [15] focuses on the characteristics of the ground truth images. Figure 9.2 depicts the DenseGAN architecture which consists of the two elements: the discriminator network [1] and the generator [15]. The fully connected convolution layer of the generator network uses the residual MSDB. Utilizing the generator network, artificial underwater images [3] are produced and the differentiation of real and synthetic data [17] is performed using the discriminator network. Two sets of convolutions, batch normalization, and LeakyReLU make up the generator network, which is followed by a pair of MSDB blocks and then extended by a deconvolution layer [18], batch normalization [20], and the LeakyReLU [19], and at the end by the deconvolution layer and tanh activation [15]. The tanh is implemented because the tanh function's gradient is four times higher than the sigmoid function's gradient. The input features are distributed in two branches in each of the several dilations that each kernel in the MSDB layer has in it. In the middle of the MSBD block, the characteristics from a single branch are combined and then passed back to the other branch. The features are once more combined and joined to a 1×1 convolution layer at the end of the MSDB block. The discriminator layer consists of a fine normalization layer in which the first and last layers contain convolution and LeakyReLU as a sequence and convolution with sigmoid for multiclass classification, respectively. The middle three layers are arranged in the order convolution, batch normalization, and LeakyReLU. As LeakyReLU allows for back-propagation even for negative input values because it has a slight positive slope in the negative area, this is implemented in this layer. This activation function aids a neural network in learning intricate linkages and data patterns. The frameworks of DenseNet and the MSDB are shown in figures 9.4(a) and (b), respectively. There are 64 and 128 feature maps [1] with kernel sizes of 7×7 and 3×3, respectively, in the first two layers of the generator. The final deconvolution layer has exactly as many channels as the input channels and the distribution of tanh is between -1 to 1. The ADAM optimizer is employed, which

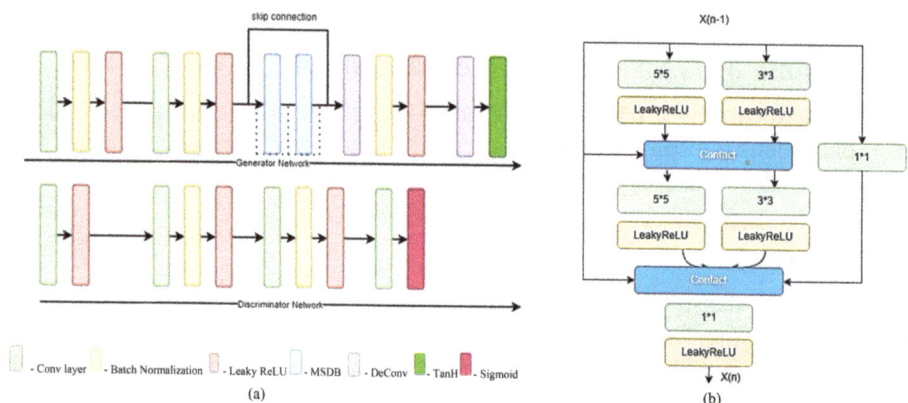

Figure 9.4. (a) The DenseNet framework architecture. 'Conv' and 'Deconv' denote the convolution layer and deconvolution layer, 'BN' denotes the batch normalization [13]. (b) MSDB framework with the skip connection [13].

has a learning rate of 10^{-3}, a batch size of 32, and patch sizes [2] of $256 \times 256 \times 3$. Without non-linear activation functions, a neural network is merely a straightforward linear regression model, however, for the final layer the linear activation function is always implemented.

9.4.2 Fusion generative adversarial network

In the fusion generative adversarial network (FGAN) [14] multiple images are passed to different sequences of layers for fusion in a single network. The architecture of FGAN is shown in figure 9.5. This is the first GAN structure that eliminates color casts, poor contrast, and haze problems using multiple image fusion as the input. The generator network comprises two convolution layers in which two inputs are passed. The first fully connected convolution network is connected to the two basic blocks and the output from the convolution layer is summed with two basic blocks and passed to the two stacks of deconvolution layers with LeakyReLU

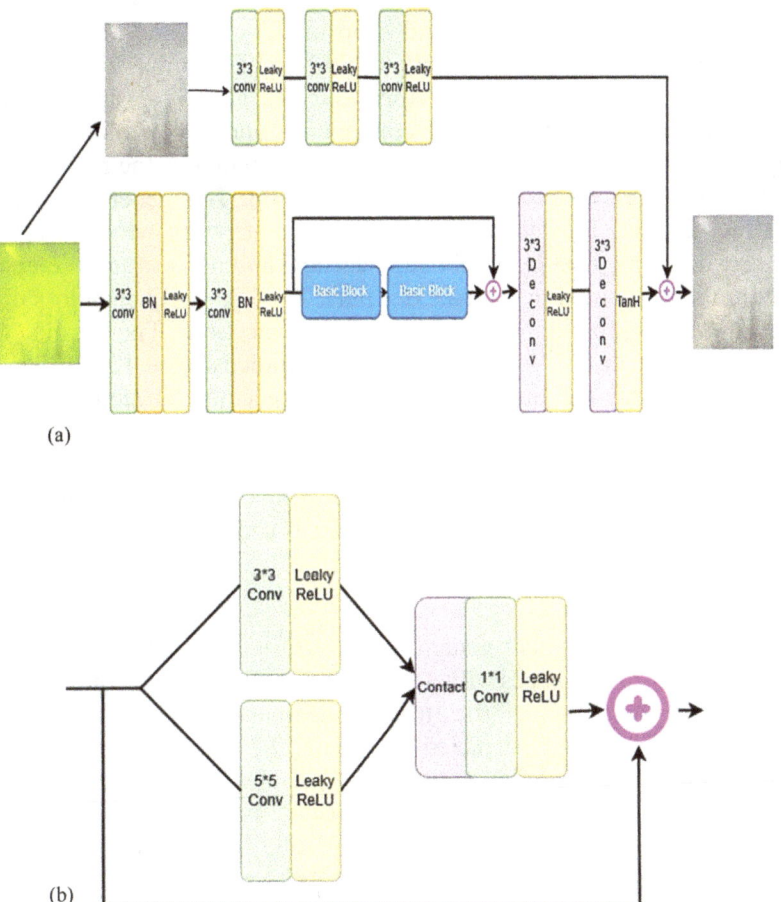

Figure 9.5. (a) The framework for FGAN [14]. (b) The building block model of FGAN [14].

and tanh activation function. The result of adding the outputs of the first convolution layer with basic blocks and the second fully linked convolution layer results in the projection of an improved image. The elements of the basic block consist of two convolution layers concatenated with another convolution layer. Combining l_2 loss, adversarial loss, and relative GAN loss creates the loss function. The input size is 256×256, the batch size is 16, and the learning rate used throughout the model is 10^{-3}. Figures 9.5 (a) and (b) illustrate the FGAN architecture and basic block structure. The spectral normalization is introduced in the discriminator network with five convolution layers. The discriminator with spectral normalization achieves better quality than that of the normal discriminator.

9.4.3 WaterGAN

WaterGAN [15] is designed to produce synthetic images that use color correction by manipulating the three RGB channels. This network is composed of two parts, of which the first part generates the synthetic data and the second part restores the underwater images with the help of two networks, color correction and depth estimation. The entire architecture and deep model of WaterGAN are shown in figures 9.6(a) and (b). The first part is WaterGAN, which comprises a generator and a discriminator. The input noise vectors that are already reshaped are passed through the generator with stacks of convolution and deconvolution layers with resulting in synthetic images. The discriminator differentiates the real and synthetic images with the result of large image data collection. This generated image along with the real datasets are fed into the restoration network which is an encoder/decoder network. The pixel-by-pixel dense learning and non-parametric upsampling processes are implemented in the encoder part. The high-frequency loss due to the pooling operation is limited by introducing skip connections. The learning rate used for depth estimation is 10^{-6} with an input image dimension of 128×128. Further, the learning rate and the image resolution in the color correction network are reduced to 10^{-7} and 512×512, respectively. The images are normalized to (0, 1) in the post-processing step and the l_2 loss function is utilized in the encoder network.

The generated images from WaterGAN are compared with other similar GANs, and the visual perception of the enhanced image is used for numerous applications.

9.4.4 UWGAN

The objective of the underwater generative adversarial network (UWGAN) is color correction in underwater images with a weakly supervised learning method. To learn the functional mapping between the atmospheric image (target) and the underwater image, the network design depicted in figure 9.7 [16] is comparable to the CycleGAN, which incorporates forward and backward networks (the source). This serves to illustrate the distinctive qualities of the source image data and the target image data for style transfer. Both the discriminator network and the generator network used in this architecture are comparable to CycleGAN. The framework is shown in figure 9.7. Cycle-consistency loss, structural similarity (SSIM) loss, and adversarial loss are added to create the final loss function.

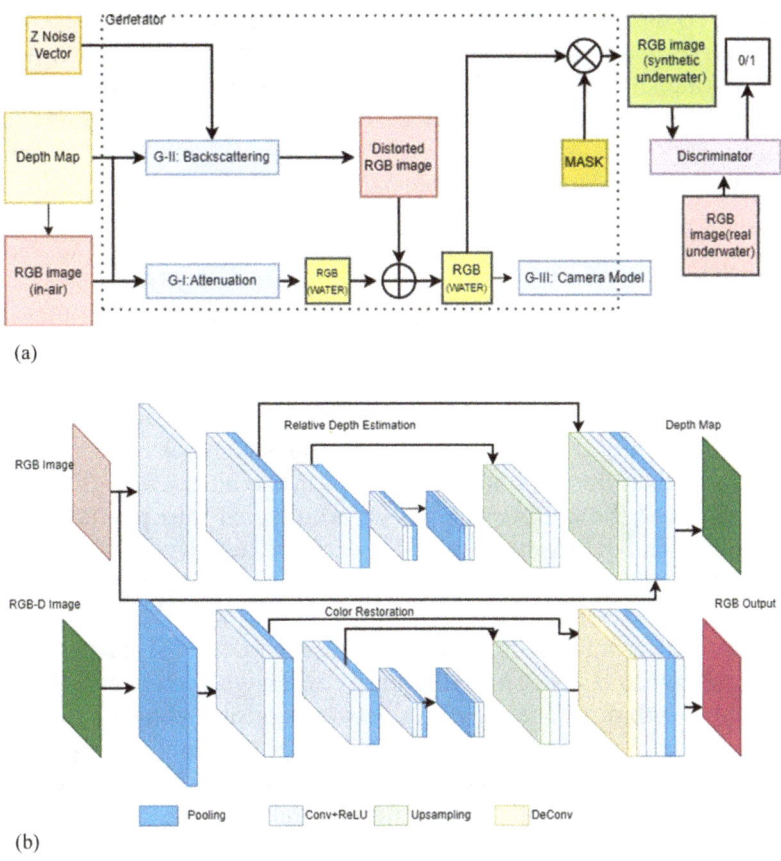

Figure 9.6. (a) Architecture of WaterGAN [15]. (b) Elaborated structure of WaterGAN model [15].

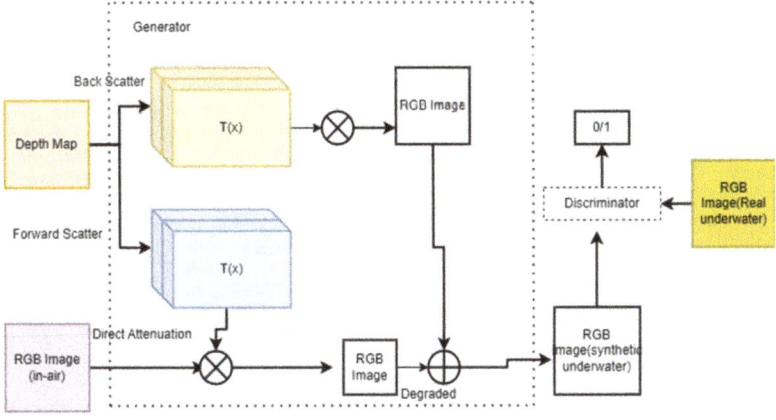

Figure 9.7. UWGAN architecture takes the depth map and RGB image as the input with consideration of forward and backward scattering elements [16].

The SSIM loss is the perceptual metric used to compare the two images. In addition to taking into account significant perceptual phenomena such as luminance and contrast masking, it views image degradation as a perceived shift in structural information. The details and structures are preserved by the SSIM loss, the distributions are matched by adversarial loss, and contradictory learning is prevented by cycle-consistency loss.

9.4.5 MCycleGAN

The multiscale cycle generative adversarial network (MCycleGAN) framework [17], as shown in figure 9.8, is a CycleGAN variant that introduces multiscale SSIM loss to improve the restoration of underwater images. The process of style transfer is used to convert from water style to air style. The transmission map of underwater images with high levels of turbidity is obtained by the dark channel prior algorithm. The depth data from the transmission map are kept in the binary filters. The high-level turbidity image and the undistorted images are divided into red, green, and blue color channels. The SSIM loss function between the high-level turbid water image and clear water image is computed by subjecting the sliding windows of different sizes into each channel. The binary filters are multiplied with the SSIM transmission map and then added together to generate the multiscale SSIM maps. The discriminator network is then fed with both the generated image and the actual underwater image with high turbidity. The generator network consists of nine ResNet modules and the discriminator consists of 70×70 PatchGANs. The input image is 256×256 in size, the ADAM optimizer is active, and the learning is fixed at a rate of 0.0002. The resultant loss function is the summation of multiscale SSIM, cycle-consistency loss, and adversarial loss.

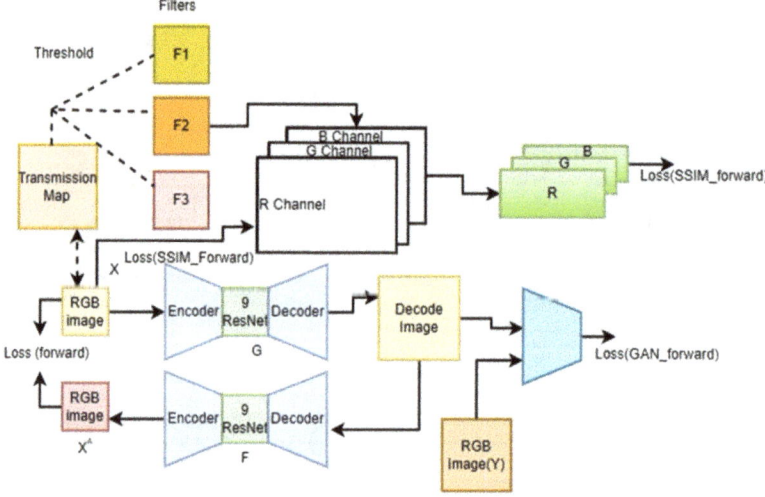

Figure 9.8. MCycleGAN framework. Here the loss functions and the mappings are denoted in arrow marks [17].

9.4.6 UIE-sGAN

Haze detection and color correction are the two modules that make up the underwater image enhancement-stacked GAN (UIE-sGAN) [18]. As seen in figure 9.9, both the generator network and the discriminator network are included in the module. The haze detection module resembles a U-Net that has seven convolution layers, a deconvolution layer added by batch normalization, and a LeakyReLU activation function in the middle layers. Batch normalization is not used in the first convolution layer, and tanh activation is used in the bottom deconvolution layer in place of LeakyReLU. Similar to how the sigmoid layer adds batch normalization and LeakyReLU to later layers, the discriminator network combines four convolution layers. The haze detecting module produces a haze mask, which is stacked on top of the color-correcting module. With the exception of the fact that the color detection modules' inputs are a haze mask and an RGB image, the architectures of the haze detection module and the color detection module are comparable. The color-corrected image is the product of the color detection module. The loss function is the combination of the consistency loss and the adversarial loss of two modules (color correction and haze detection), and the input image has a dimension of 256×256. The optimizer used is ADAM, and the learning rate is fixed as 5×10^{-5}.

9.4.7 UGAN

The underwater generative adversarial network (UGAN) [19], as shown in figure 9.10, is introduced to enhance the quality of UWIs. In the discriminator network, the Wasserstein GAN gradient penalty is used and batch normalization is not implemented to reduce the computation level. The generator network and

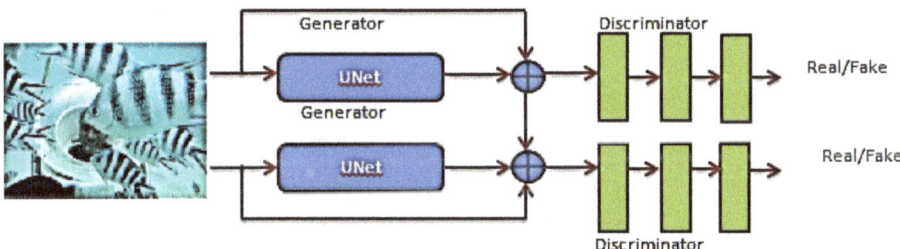

Figure 9.9. The UIE-sGAN framework with the generator module and the discriminator module [18].

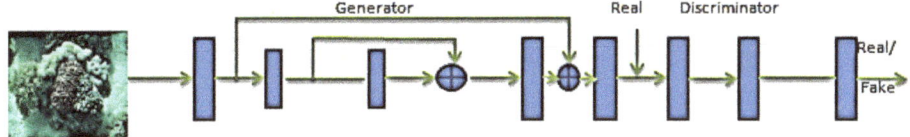

Figure 9.10. The UGAN architecture. The first five symbols represent the generator convolution layers and the last three blocks represent the discriminator module [19].

discriminator network are the same as for CycleGAN and PatchGAN, respectively. In this case, the distortion model used to create the paired images is the CycleGAN. As it learns the style of distorted images and undistorted images in the underwater environment, the CycleGAN is meant to carry out the style transfer. The optimizer used is the ADAM optimizer to perform tuning by using fewer parameters. The input image size is 256 × 256, the rate of learning is fixed, that is 10–4, and the summation of l_1 loss, gradient descent loss, and the Wasserstein-1 loss results in the model loss function. The goal of the Wasserstein loss function is to widen the disparity between the ratings for actual and synthetic images.

9.4.8 AquaGAN

The mapping between the generated image and the ground truth image will be learned using the AquaGAN [20] architecture, as shown in figure 9.11. Similar to other GANs, this architecture also includes a couple of encoders and a decoder. The discriminator network is used to improve the degraded underwater image. The encoder learns the absorption and scattering factors to improve the image quality. The input underwater image with a high degradation level (D) is fed into the top encoder E_θ^a and the latent representation of D, X_c is of the dimension $M \times 1$. The input to bottom encoder E_θ^b is the approximated Jerloy patch that is used to generate the latent space of dimension $M \times 1$. The framework of the AquaGAN is shown in figure 9.11. The correlation between the two encoders is concatenated to create the single ($2M \times 1$) vector. To recreate the underwater image, the decoder obtains the vector with the dimensions ($2M \times 1$). The attenuation coefficient is taken into account for restoring the underwater image. The sum of content loss and style loss is the AquaGAN model loss function.

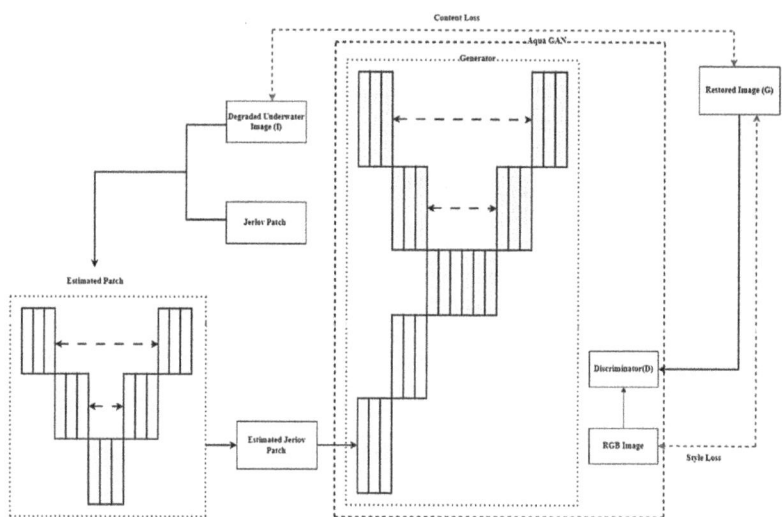

Figure 9.11. AquaGAN framework. Here the route of the input and the loss functions are represented using the arrows [20].

9.5 Underwater image datasets

The architectures discussed in this chapter are trained using different benchmark datasets. In underwater visual perception and image/video processing, it is crucial to achieve quality assessment in terms of both subjective and objective perspective for underwater images. Table 9.1 lists the several dataset types that are frequently employed for underwater image improvement and restoration. These datasets have been implemented in recent research works for UWI analysis. Each dataset is formulated in a different underwater environment and is used in a variety of applications. These datasets serve as the primary source for producing several synthetic images.

Table 9.1. Various publicly available benchmark datasets for underwater image enhancement.

Datasets	Description	Number of images
UIEB dataset [23, 24]	Underwater images in different lighting conditions.	890 images (630 MB +786 MB) size
Enhanced Underwater Visual Perception (EUVP) [21, 22]	Paired data, underwater dark, underwater ImageNet, underwater scenes, unpaired data, various lighting conditions.	10 K 25 K
UFO-120 dataset [25, 26]	Dataset collected in different types of ocean water.	1620
Ocean Dark dataset [35]	Different images at varied depth levels.	183
Synthetic Underwater Image dataset (SUID) [27, 28]	Contains ground truth images and synthetic underwater images.	30 ground truth and 900 generated underwater images (56.87 MB)
Real-world Underwater Image Enhancement (RUIE) [29, 30]	Images captured underwater taken using natural light and categorized into three distinct groups or sets.	4230
Sea-thru [34]	Images captured using a pair of stereo cameras.	1157
TURBID [33]	Images were captured inside a water tank and divided into five subsets based on the level of turbidity, or the number of particles or impurities present in the water that affect its clarity.	112
Large Scale Underwater Image dataset (LSUI) [36]	Images captured under varying lighting conditions and levels of cloudiness and taken for different subjects.	5004
U-45 [31, 32]	Images that have varying degrees of contrast, low quality, and appear hazy because of several factors.	45

9.6 Performance standards

All designed networks for pre-processing are evaluated by some standard metrics which can be objective or subjective analysis. These standards are termed quality evaluation metrics. Reference quality metrics and non-reference quality metrics are additional categories. Non-reference metrics are those that quantify quality without directly comparing the raw and treated versions of images.In contrast to this, reference metrics refer to quality measured through a direct comparison of the raw and modified images. The evaluation metrics for analysing image quality are shown in figure 9.12. The most commonly used reference metrics are peak signal to noise ratio (PSNR), SSIM, and mean square error (MSE). As it is a challenging task to acquire reference images, the quality metrics can be measured without reference images in the underwater environment. Some of the non-reference metrics are as follows:

- Underwater image quality measure (UIQM) [37] consists of three attributes, colorfulness (underwater image colorfulness measure) [38], sharpness (underwater image sharpness measure), and contrast (underwater image contrast measure) [39].
- Patch based contrast quality index (PCQI) [40] depends on mean intensity, structural distortion, and change in contrast of an image.
- The underwater color image quality evaluation metric (UCIQE) [41] combines blurriness, low contrast, and color cast for the evaluation.

Image entropy, contrast code image (CCI) [42], and image quality index are other non-reference metrics that are used commonly for quality measurement [17].

To obtain a good quality result the MSE value must always be small and the other evaluation metric values must be higher. The UWI quality evaluation parameters are listed in table 9.2. The various categories of UWI can be evaluated for a particular algorithm to understand its functionality. Based on the values of estimated parameters, a particular algorithm can be used for different applications [10].

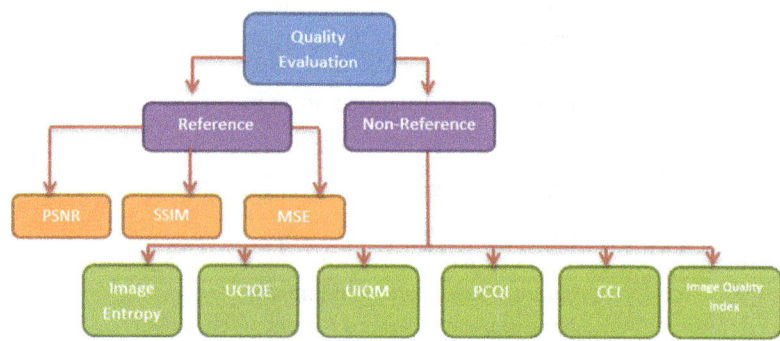

Figure 9.12. Various quality evaluation metrics.

Table 9.2. Metrics for underwater image enhancement.

Metrics	Equation	Parameters
Mean square error (MSE)	$\frac{1}{XY}\sum_{i=1}^{X}\sum_{j=1}^{Y}[K(i,j) - L(i,j)]^2$	$K(i,j)$ original image, $L(i,j)$ enhanced image, $X \times Y$ size of image
Peak signal to noise ratio (PSNR)	$20\log_{10}\left(\frac{\max}{\sqrt{MSE}}\right)$	max is the maximum pixel
Structure similarity index measure (SSIM)	$\frac{(2\mu_x\mu_y + C_1)(2\sigma_{xy}+C_2)}{(\mu_x^2\mu_y^2+C_1)(\sigma_x^2\sigma_y^2+C_2)}$	μ_x, μ_y mean value, σ_x, σ_y standard deviation, patch x, y
Underwater color image quality evaluation (UCIQE)	$C_2 \times contrast_l + C_1 \times \sigma_{chroma} + C_3 \times \mu_{saturation}$	C_1, C_2, C_3 weight coefficient
Underwater image quality measure (UIQM)	$Coeff_1 \times UICM + Coeff_2 \times UISM + Coeff_3 \times UIConM$	UICM, UISM, and UIConM, colourfulness, sharpness, and contrast measures, respectively
Patch based contrast quality index (PCQI)	$\frac{1}{P}\sum_{k=1}^{P} l_q(m_i, j_i) l_r(m_i, j_i) l_s(m_i, j_i)$	P patches, l_q, l_r, l_s comparison functions

9.7 Future scope

Underwater image processing (UIP) is an interesting research domain and has improved a lot in the past few years. The rapid advancement of deep learning techniques is one of the primary drivers of growth. But there are also some areas that need improvement, such as dehazing, restoration, resolution, deblurring, and color correction [4]. Here, the list of areas to concentrate more on in the future is discussed. A lot of methods generate synthetic data as there is a lack of datasets of UWIs for training models. Even though these models provide the best result for synthetic datasets they fail on real UWIs. Thus there is always a need to develop transfer learning approaches that can work with only a few available images and with fewer computations [7].

As there are limited datasets researchers can focus on generating synthetic datasets. The quality of UWIs can be improved further by combining all the parameters, such as light illumination, color correction, and hazy effects [28]. Future researchers can concentrate on different color spaces to improve the clarity of images and review pre-processing techniques based on color space from traditional to recent advancements.

Even though there are many advantages to using deep networks, there are also a few drawbacks, as large images need to be considered and hence the computations are very complex. In UIP the ground truth image is also synthetic data which greatly affects the evaluation metrics.

9.8 Conclusion

UIP is a fascinating field for researchers to explore new concepts and develop new models. The GAN is emerging as an effective technique for the generation of synthetic data with the help of latent space in a random fashion. The key feature of GAN based model is that detailed accounting is not necessary for high-level mathematical computations. This has motivated the research community to use GANs in various applications. This chapter summarizes the basic framework of the GAN, variants of GAN for image enhancement and restoration in underwater environments, the structure of GANs that have been developed by various authors for underwater images, the several publicly available datasets for enhancement, and the classification of evaluation metrics with formulation for evaluating the result. The significant surge in GAN-based models stems from their ability to generate realistic synthetic data by utilizing large amounts of unlabeled data through adversarial training. There is immense scope in the development of architectures and algorithms that could be applied in various fields. The field of deep learning has grown significantly with the help of GAN, and its development process is still ongoing.

References

[1] Han M, Lyu Z, Qiu T and Xu M 2018 A review on intelligence dehazing and color restoration for underwater images *IEEE Trans. Syst., Man, Cybernet.: Syst.* **50** 1820–32
[2] Treibitz T and Schechner Y Y 2012 Turbid scene enhancement using multi-directional illumination fusion *IEEE Trans. Image Process.* **21** 4662–7

[3] Peng Y T and Cosman P C 2017 Underwater image restoration based on image blurriness and light absorption *IEEE Trans. Image Process.* **26** 1579–94

[4] Liu P, Wang G, Qi H, Zhang C, Zheng H and Yu Z 2019 Underwater image enhancement with a deep residual framework *IEEE Access* **7** 94614–29

[5] Cai B, Xu X, Jia K, Qing C and Tao D 2016 DehazeNet: an end-to-end system for single image haze removal *IEEE Trans. Image Process.* **25** 5187–98

[6] Zhu J Y, Park T, Isola P and Efros A A 2017 Unpaired image-to-image translation using cycle-consistent adversarial networks *Proc. IEEE Int. Conf. on Computer Vision* pp 2223–32

[7] Tan R T 2008 Visibility in bad weather from a single image *2008 IEEE Conf. on Computer Vision and Pattern Recognition* (Piscataway, NJ: IEEE) pp 1–8

[8] Lu J, Yuan F, Yang W and Cheng E 2021 An imaging information estimation network for underwater image color restoration *IEEE J. Ocean. Eng.* **46** 1228–39

[9] Xu L and Wei Y 2022 'Pyramid Deep dehazing': an unsupervised single image dehazing method using deep image prior *Opt. Laser Technol.* **148** 107788

[10] Li C, Guo J and Guo C 2018 Emerging from water: underwater image color correction based on weakly supervised color transfer *IEEE Signal Process Lett.* **25** 323–7

[11] Panetta K, Kezebou L, Oludare V and Agaian S 2021 Comprehensive underwater object tracking benchmark dataset and underwater image enhancement with GAN *IEEE J. Ocean. Eng.* **47** 59–75

[12] Goodfellow I J, Pouget-Abadie J, Mirza M, Xu B, Warde-Farley D, Ozair S, Courville A and Bengio Y 2014 Generative adversarial nets *Proc. 27th Int. Conf. on Neural Information Processing Systems* vol 2 pp 2672–80

[13] Guo Y, Li H and Zhuang P 2019 Underwater image enhancement using a multiscale dense generative adversarial network *IEEE J. Ocean. Eng.* **45** 862–70

[14] Li H, Li J and Wang W 2019 A fusion adversarial underwater image enhancement network with a public test dataset arXiv:1906.06819

[15] Li J, Skinner K A, Eustice R M and Johnson-Roberson M 2017 WaterGAN: unsupervised generative network to enable real-time color correction of monocular underwater images *IEEE Robot. Autom. Lett.* **3** 387–94

[16] Wang N, Zhou Y, Han F, Zhu H and Yao J 2019 UWGAN: underwater GAN for real-world underwater color restoration and dehazing arXiv:1912.10269

[17] Lu J, Li N, Zhang S, Yu Z, Zheng H and Zheng B 2019 Multi-scale adversarial network for underwater image restoration *Opt. Laser Technol.* **110** 105–13

[18] Lane H C and D'Mello S K 2019 Uses of physiological monitoring in intelligent learning environments: a review of research, evidence, and technologies *Mind, Brain and Technology: Learning in the Age of Emerging Technologies* (Springer) pp 67–86

[19] Fabbri C, Islam M J and Sattar J 2018 Enhancing underwater imagery using generative adversarial networks *IEEE Int. Conf. on Robotics and Automation (ICRA)* (Piscataway, NJ: IEEE) pp 7159–65

[20] Desai C, Reddy B S S, Tabib R A, Patil U and Mudenagudi U 2022 AquaGAN: restoration of underwater images *Proc. IEEE/CVF Conf. on Computer Vision and Pattern Recognition* pp 296–304

[21] Badrinarayanan V, Kendall A and Cipolla R 2017 SegNet: a deep convolutional encoder-decoder architecture for image segmentation *IEEE Trans. Pattern Anal. Mach. Intell.* **39** 2481–95

[22] Pizarro O, Friedman A, Bryson M, Williams S B and Madin J 2017 A simple, fast, and repeatable survey method for underwater visual 3D benthic mapping and monitoring *Ecol. Evol.* **7** 1770–82

[23] Janoch A, Karayev S, Jia Y, Barron J T, Fritz M, Saenko K and Darrell T 2013 A category-level 3D object dataset: putting the kinect to work *Consumer Depth Cameras for Computer Vision: Research Topics and Applications* (Springer) pp 141–65

[24] Lai K, Bo L and Fox D 2014 Unsupervised feature learning for 3D scene labeling *IEEE Int. Conf. on Robotics and Automation (ICRA)* (Piscataway, NJ: IEEE) pp 3050–7

[25] Shotton J, Glocker B, Zach C, Izadi S, Criminisi A and Fitzgibbon A 2013 Scene coordinate regression forests for camera relocalization in RGB-D images *Proc. IEEE Conf. on Computer Vision and Pattern Recognition* pp 2930–7

[26] Goodfellow I, Pouget-Abadie J, Mirza M, Xu B, Warde-Farley D, Ozair S, Courville A and Bengio Y 2020 Generative adversarial networks *Commun. ACM* **63** 139–44

[27] Johnson J, Alahi A and Fei-Fei L 2016 Perceptual losses for real-time style transfer and super-resolution *Proc. 14th European Conf. on Computer Vision* vol 14 *(Amsterdam, 11–14 October)* (Berlin: Springer) pp 694–711

[28] Anwar S and Barnes N 2020 Densely residual Laplacian super-resolution *IEEE Trans. Pattern Anal. Mach. Intell.* **44** 1192–204

[29] Berman D, Levy D, Avidan S and Treibitz T 2020 Underwater single image color restoration using haze-lines and a new quantitative dataset *IEEE Trans. Pattern Anal. Mach. Intell.* **43** 2822–37

[30] Skinner K A and Johnson-Roberson M 2017 Underwater image dehazing with a light field camera *Proc. IEEE Conf. on Computer Vision and Pattern Recognition Workshops* pp 62–9

[31] Yang M and Sowmya A 2015 An underwater color image quality evaluation metric *IEEE Trans. Image Process.* **24** 6062–71

[32] Duarte A, Codevilla F, Gaya J D O and Botelho S S 2016 A dataset to evaluate underwater image restoration methods *OCEANS 2016-Shanghai* (Piscataway, NJ: IEEE) pp 1–6

[33] Akkaynak D and Treibitz T 2019 Sea-Thru: a method for removing water from underwater images *Proc. IEEE/CVF Conf. on Computer Vision and Pattern Recognition* pp 1682–91

[34] Porto Marques T, Branzan Albu A and Hoeberechts M 2019 A contrast-guided approach for the enhancement of low-lighting underwater images *J. Imag.* **5** 79

[35] Peng L, Zhu C and Bian L 2023 U-shape transformer for underwater image enhancement *Workshops: Proc. Computer Vision–ECCV 2022, Part II (Tel Aviv, 23–27 October)* (Cham: Springer) pp 290–307

[36] Naik A, Swarnakar A and Mittal K 2021 Shallow-UWnet: compressed model for underwater image enhancement (student abstract) *Proc. AAAI Conf. on Artificial Intelligence* vol 35 pp 15853–4

[37] Panetta K, Gao C and Agaian S 2015 Human-visual-system-inspired underwater image quality measures *IEEE J. Ocean. Eng.* **41** 541–51

[38] He K, Zhang X, Ren S and Sun J 2016 Deep residual learning for image recognition *Proc. IEEE Conf. on Computer Vision and Pattern Recognition* pp 770–8

[39] Ledig C et al 2017 Photo-realistic single image super-resolution using a generative adversarial network *Proc. IEEE Conf. on Computer Vision and Pattern Recognition* pp 4681–90

[40] Isola P, Zhu J Y, Zhou T and Efros A A 2017 Image-to-image translation with conditional adversarial networks *Proc. IEEE Conf. on Computer Vision and Pattern Recognition* pp 1125–34

[41] He K, Sun J and Tang X 2010 Single image haze removal using dark channel prior *IEEE Trans. Pattern Anal. Mach. Intell.* **33** 2341–53

[42] Ye X, Xu H, Ji X and Xu R 2018 Underwater image enhancement using stacked generative adversarial networks *Proc. 19th Pacific-Rim Conf. on Advances in Multimedia Information Processing–PCM 2018 (Hefei, China, 21–22 September)* (Cham: Springer) pp 514–24

[43] Islam M J, Xia Y and Sattar J 2020 Fast underwater image enhancement for improved visual perception *IEEE Robot. Autom. Lett.* **5** 3227–34

IOP Publishing

Data Analytics for Intelligent Systems
Techniques and solutions
Sachin Taran, Chhavi Dhiman and Manjeet Kumar

Chapter 10

Leveraging knowledge graphs for the analysis and recommendation of jobs

Amit Patil, Muskan Jain, Neetu Sardana, Sanjoli Goyal and Deepika Varshney

One of the most important aspects of the modern recruitment market is online job boards. There is a high demand for reliable, effective, and clear job recommendations due to the large number of people searching for jobs on the Internet. Although there has been progress in the use of recommender systems in various online fields, there is still room for improvement in the area of job recommendations. Job seekers often spend a significant amount of time sifting through a large amount of information on the Internet to find the most relevant and helpful job opportunities. We propose a bipartite graph-based job-hunting recommendation system. The system is efficient in comparison to existing collaborative filtering as it handles spareness and helps in quick decision-making. The chapter also analysed the resumes and job descriptions separately using homogeneous graphs. The analysis has been carried out using varied features such as age, domain, skills, and gender.

10.1 Introduction

The growing popularity of the Internet has increased the demand for online job searching. Job sites enable job seekers to build and enhance network opportunities. According to Jobsite's report in 2014, 68% of online job seekers were college graduates or postgraduates, but now the scenario has changed.

Post-pandemic, many companies have begun firing people, for instance recently Facebook parent Meta laid off 11 000 people, about 13% of its workforce [1]. Twitter (now X) has fired about 50 percent of its workforce, which is around 3800 employees. In addition, the post-pandemic 'new normal' has led to a rise in demand for diverse jobs. Many people of varied backgrounds and experience are looking for the right job on online platforms [1].

The main issue for most job-searching websites is that they simply present recruitment information to website visitors. Job seekers manually gather the necessary information to locate and apply for jobs. They have to apply their wisdom to analyse the information presented on the job sites. The entire process is time-consuming and inefficient. An efficient job recommender system is required that can help the job hunter to make quick and timely decisions

Generally, past studies have used collaborative filtering (CF) approaches [2] or a content-based (CB) approach for recommendations [3]. Collaborative filtering (CF) based recommender systems use data about a user's past actions or preferences to identify items (such as jobs) that may be of interest to them. The idea behind this approach is that people who have similar interests or preferences may also like similar items. On the other hand, content-based (CB) recommender systems use information about the characteristics or attributes of items (such as job descriptions) and compare them to a user's preferences (such as their resume) to recommend items that match the user's interests. Both CF and CB approaches can be used to support users during the decision-making process by suggesting relevant items that may be of interest to them. CF techniques only use previous user behavior (such as ratings and purchases) to suggest items. However, the temporary nature of the items (jobs) in the system and the high rate at which new users and jobs are added create a significant issue called the cold start, which hinders the effectiveness of collaborative filtering techniques.

Graph-based models [12, 13] use techniques from graph theory to overcome the limitations of CF approaches, such as sparsity, and enhance the accuracy of recommendations. Link analysis allows us to compute centrality measures such as degree, betweenness, closeness, and eigenvector to visualize the connections on a link chart or link map. Graph-based recommendation systems [14, 15] are classified based on how they construct the graph and use it to make recommendations. We employed heterogeneous graph-based models to create a bipartite graph of resumes and job descriptions. They are connected with a common node which can be 'Skill', 'DevType', 'Age', etc. Using a bipartite graph, we have performed job recommendations using two techniques, Adamic-Adar and common neighbors. We have also investigated homogeneous graph models for resumes and job descriptions in isolation. The work in this chapter has two dimensions. First, we have analysed the data present in the resume and job description separately. We constructed a homogeneous graph and performed analysis on resumes and job descriptions and addressed following six questions:

  (a) What are the hot and cold skills in the available resume and job description?
  (b) What is the distribution of resumes gender-wise corresponding to different ages?
  (c) What is the distribution of salaries corresponding to different age groups?
  (d) What is the correlation between domain and skills?
  (e) What are the top skills used in various domains in the resume?
  (f) Describe the jobs with similar skills as a resume.

In the second dimension, we constructed a heterogeneous bipartite graph between the resume and job description and performed the job recommendations.

The rest of the chapter is organized as follows. Section 10.2 presents past studies based on recommendations, section 10.3 describes the proposed methodology, section 10.4 provides the experimental details, and the chapter ends with a conclusion.

The chapter includes several figures that illustrate the proposed framework of job recommendation using knowledge graphs (KGs). Figure 10.1 presents an overview of the proposed framework, while figures 10.2 and 10.3 show the homogeneous graphs of resumes and job descriptions, respectively. Figures 10.4 and 10.5 provide a bar graph of the percentage of skills and domains present in the resume and job description datasets. Figures 10.6 and 10.7 provide bar graphs of the percentage of skills and domains present in the job description dataset. Figure 10.8 presents a heat map to correlate domain with skills. Figure 10.9 provides a bar graph of the top three skills worked with a particular domain. Figure 10.10 shows the list of jobs that have similar skills to a particular resume. Figures 10.11(a) and (b), 10.12(a) and (b), 10.13(a) and (b), and 10.14(a) and (b) present graphs and scores between a resume and different jobs, the relationship between a particular resume and job, an ordered list of the common neighbor score and Adamic-Adar score, and graphical representation of the relationship between resume and job and scores, respectively.

10.2 Related work

The various studies [4–6, 9, 11] closely related to our work are presented in this section.

The study in [4] introduced a new method for developing a job recommendation system designed to be scalable and reliable for online recruiting services. This approach involves creating a multigraph (a type of graph with multiple edges between the same nodes) of jobs that are connected by similar edges based on user behavior and job content. The study found that the recommendations generated by this system had an average accuracy of around 90% and performed better than a traditional collaborative filtering approach. In addition, the system required only a third of the number of emails to be sent to produce the same level of performance. This suggests that the proposed method may be an effective and efficient way to recommend jobs to users in online recruiting services [4].

Different methods discussed in previous studies explained the development phases of the recommender system. First, deep learning models are used widely to extract latent features from item content and auxiliary information, improve recommendation accuracy, and handle sparse data. Autoencoders and CNNs are the most commonly used models for this purpose. Second, a recommender system needs to incorporate data from multiple sources to obtain accurate user preferences and intentions. Third, the user profile is a critical component of the recommendation process. Vector-based profiles, which model users as vectors of numeric ratings, are the most commonly used. However, these profiles can contain sensitive information, so histogram-based profiles, which model user data as samples of predefined categories, are sometimes used to protect privacy. Fourth, the performance and quality of a recommender system can be evaluated using various metrics such as Mean Square Error (MSE) , Mean Absolute Error (MAE) and Receiver Operating

Characteristics (ROC) for prediction accuracy and top-N recommendations. Techniques such as generalization and deep learning can be used as preprocessing steps in the development of a recommendation system to improve its performance and reduce the dimensionality of large datasets [5].

Several platforms and systems have been developed to help people find jobs, including GUapp, which is a platform that assists users in finding jobs in the Italian public sector [6]. GUapp uses a technique called latent Dirichlet allocation to recommend jobs to users based on their skills and preferences. In addition to this, the platform also includes a chatbot that allows users to communicate with the system using natural language to refine their job search. This chatbot feature makes the job search process more interactive and allows users to adjust their search criteria as they go along. These types of platforms and systems can be useful for helping people find job opportunities that are relevant to their skills and interests [6].

10.3 Proposed methodology

Figure 10.1 presents an overview of the proposed framework for a job recommender using a knowledge graph (JRKG). The various stages in the framework are described below.

Data collection. For this study, we have used two datasets. The first dataset consists of resumes of the job hunters and another dataset contains the job descriptions (JDs) of the available jobs. The resume dataset is extracted from the Kaggle website. We have considered seven features of a particular resume. The job description dataset was extracted from USjobs on Dice.com.

Data cleaning and preprocessing. In this stage the features in both datasets are cleaned and preprocessed. The 'Salary' field values were either weekly, monthly, or annual, we used the 'salary_type' field to achieve the salary as a discrete value depicting the annual salary. Since the salary field had discrete values we performed binning to represent it with uniformity. The four attributes 'LanguageWorkedWith', 'DatabaseWorkedWith', 'PlatformWorkedWith', and 'FrameworkWorkedWith' were eliminated from the table, and a single attribute namely 'Skills' has been created which collectively contains the data values of these four attributes.

Data analysis and visualization. The features in the resumes and JDs are analysed in varied dimensions. Hot and cold skills have been investigated to analyse the skill set in demand. The chapter also analysed the job seeker's gender, domain, and age. We also analysed whether there is any variation in the salary slabs corresponding to age or gender. Various plots such as histograms, bar graphs, heat maps, and scatter plots have been used for data analysis.

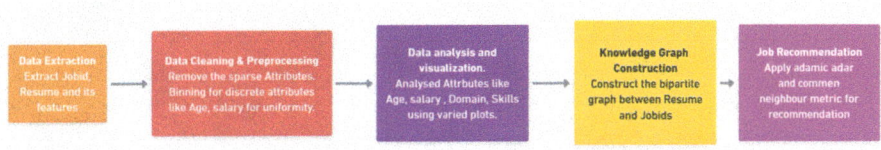

Figure 10.1. The proposed framework of the job recommender using a knowledge graph.

Knowledge graph construction. To construct knowledge graphs, we considered nodes of three types, namely 'Resume', 'Skill', and 'JD_id'. Each resume node has a link with its skills set and, similarly, JD_ids are linked to the skills required for that job. To correlate and connect the resumes and JDs and represent this in the form of a knowledge graph, common skills and domains are used. Figures 10.2 and 10.3 show a glimpse of what the constructed knowledge graphs would look like.

Job recommendation. A knowledge graph is used to predict the suitable job (JD_id node) for the job seekers (Resume node). The following process is used for recommendation:

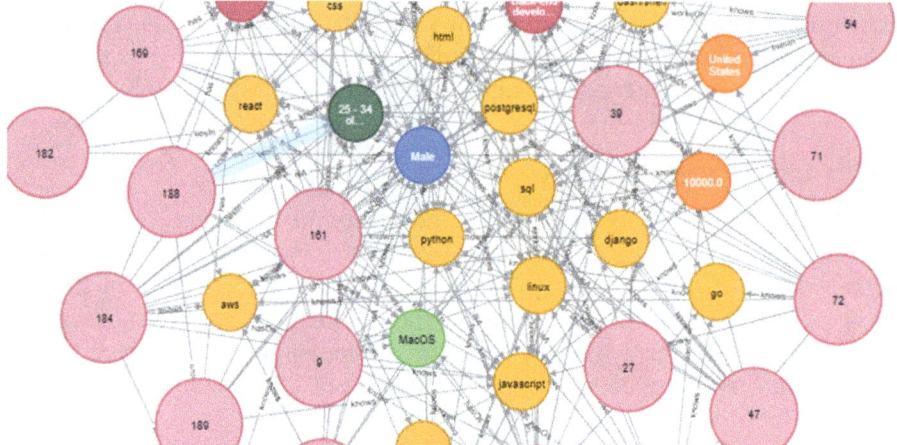

Figure 10.2. Homogeneous graph of the resume dataset.

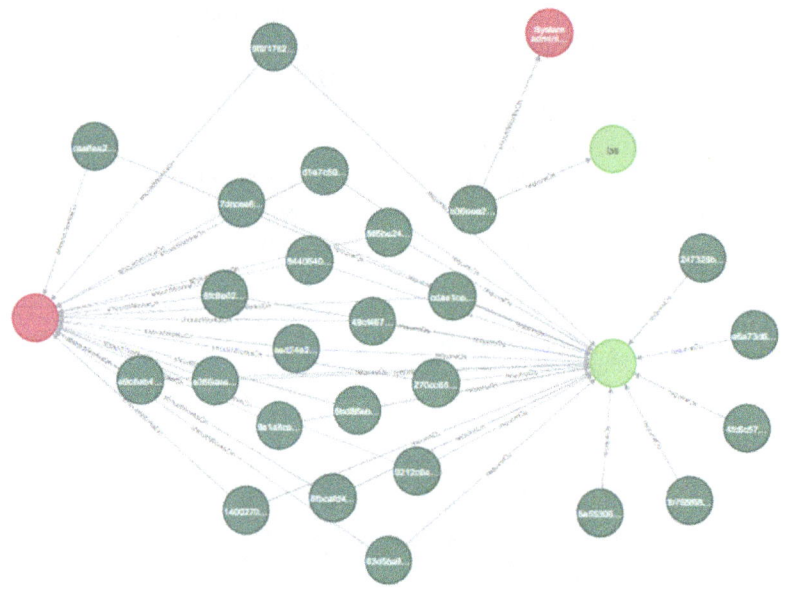

Figure 10.3. Homogeneous graph of the job description dataset.

1. Matching the skills and domains of a ResumeID node with all the JobID nodes. That is, when we are given a resume, we derive the skills it has to offer and search for those skills within our KG. Then those skills are matched with the JobIDs already present in our database. So, transitively, a resume is connected to a JobID via a common connecting link, which is skills and domain.

2. Counting the number of matches between a resume and a JobID. That is, we now count how many connections a resume forms with a particular JobID. Count here refers to the number of intermediary nodes between the ResumeID and JobID. Note that we always use the same resume with different JobIDs because we are trying to recommend a job to a job seeker.

3. Calculating the common skill and common domain. Among the counts, there are two types of connecting nodes: the skill property node and a domain property node. We count them separately as both of them might be forming an equal number of connections between the ResumeID and JobID, but the percentage significance that they hold when we seek a job is very different.

4. Choosing the resultant JobID with a minimum of 10% skills in common with ResumeID. To further refine our recommendation we only take those JobIDs to the next step which have at least 10% of skills in common with the total skills offered by the ResumeID. We remove those JobIDs from our list of potential jobs that the job seeker would be interested in if only less than 10% of skills are all that they have in common. This also corresponds to real-life situations where both a job seeker and the employer would like to have common interests in terms of what they have to work with.

5. Taking the scores of Adamic-Adar and common neighbor between a particular resume and the jobs present in the database, if both suggest the same job then we take it to the next step.

6. Ordered by this score (in descending order) we select the first few recommendations and called this the 'JobRecommended list'.

10.4 Experimental results

10.4.1 Dataset used

We have used two datasets. One dataset consists of resumes of users and the second dataset consists of job descriptions for which jobs are available.

- The user resume dataset that is being referred to in this research was provided by Stack Overflow on the Kaggle website in 2018. Stack Overflow conducted a survey in which they asked developers about their preferences, technologies, and job preferences to create their resumes. There are 98 855 responses in this publicly available dataset [7]. The dataset includes information about the respondent, their country, company size, dependents, type, gender, job satisfaction, last new job, salary, salary type, programming languages worked with, databases worked with, platforms worked with, frameworks worked with, operating systems, and age.

- The job description dataset that is being referred to was created by PromptCloud's web-crawling service. This dataset is a subset of a larger dataset containing over 4.6 million job listings, which was created by extracting data from Dice, a US-based technology job board, in 2017. There are 22 000 job profiles in this publicly available dataset [8]. The dataset includes information about the job ID, skills required, domain, platform, and databases used for the job.

10.4.2 Evaluation parameters

We used three metrics in this chapter for evaluation, namely support for mining the skills in resumes and Adamic-Adar for a job recommendation.

The concept of support (Sp) is being used to identify the most frequent skills in a set of resumes. Support is a measure of how often a particular skill appears in a resume, and it is calculated by dividing the number of resumes that have a given skill (PS) by the total number of resumes (n) in a particular category. In other words, support measures the fraction of resumes that include a particular skill. Support can be calculated using

$$\text{Support (Sp)} = \frac{\sigma(\text{PS})}{n}. \tag{10.1}$$

Adamic-Adar score. It is a measure that is used to compute the closeness of nodes (such as job seekers and job opportunities) based on the number of shared neighbors they have. This metric is often used in social networks to predict future links and is calculated using the following formula:

$$A(x, y) = \sum_{u \in N(x) \cap N(y)} \frac{1}{\log \ | N(u)|}, \tag{10.2}$$

where $N(u)$ is the degree (number of neighbors) of each shared neighbor.

A value of 0 indicates that the nodes are not close to each other, while higher values indicate that the nodes are closer.

We have verified results using the Adamic-Adar closeness score which gives a non-zero score value while applying link prediction which means that the recommended job description is close to the resume.

Common neighbors.

$$CN(x, y) = \ | N(x) \cap N(y)|, \tag{10.3}$$

where $N(x)$ is the set of nodes adjacent to node x, and $N(y)$ is the set of nodes adjacent to node y. A value of 0 indicates that two nodes are not close, while higher values indicate nodes are closer.

The library contains a function to calculate the closeness between two nodes.

10.4.3 Tools used for evaluation and benefits

Adamic-Adar score. One way to validate the results of a link prediction algorithm is to compare the Adamic-Adar scores of the predicted links to those of existing links in the network. Higher Adamic-Adar scores for predicted links compared to existing links may suggest that they are more likely to be influential within the network. It is also possible to compare the Adamic-Adar scores of the predicted links to those produced by other link prediction algorithms. Consistently higher Adamic-Adar scores for the predicted links compared to other algorithms may indicate that the link prediction algorithm being used is particularly effective at identifying important or influential links within the network.

A positive, non-zero Adamic-Adar score can be an indication that the predicted relationship between two nodes in the network is meaningful and potentially important. This is because Adamic-Adar is a measure of the similarity between two nodes based on the number and importance of their common neighbors. A positive, non-zero score suggests that the two nodes have at least one common neighbor and that this common neighbor is not highly connected (i.e. not a 'popular' node). This can be interpreted as an indication that the common neighbor is contributing meaningful information about the relationship between the two nodes.

10.4.4 Experimental results

Data were analysed in two stages. During the first stage, data were visualized and the five questions listed in the introduction were addressed. In the second stage, a heterogeneous bipartite graph was constructed between the resumes and jobs for a recommendation.

Q1. What are the hot and cold skills in available CVs?

Motivation: We examined popular and rare skills in resumes and available job descriptions to find the level of proximity among them

Results and inferences: We applied association rule mining techniques, the Apriori and FP Growth algorithms to find the frequent skills. We considered support to be 50%. Popular skills and rare skills from the resume dataset are shown in figure 10.4.

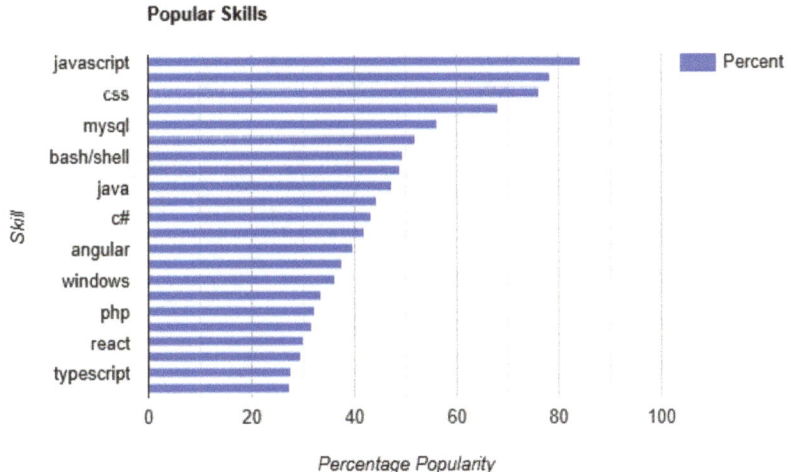

Figure 10.4. Bar graph of percentage skills present in the resume dataset.

Popular_skills of a threshold above 50%:
['node.js', 'mysql', 'sql', 'css', 'html', 'javascript']
Rare_skills of a threshold below 3%:
['predix', 'hack', 'julia', 'ocaml', 'cobol', 'mainframe', 'gaming console', 'torch/ pytorch', 'erlang', 'google home', 'apple watch or apple tv', 'ibm cloud or watson', 'apache hbase', 'clojure', 'f#', 'esp8266', 'delphi/object pascal', 'apache hive', 'haskell', 'google bigquery', 'rust', 'ibm db2', **'neo4j'**, 'windows phone', 'salesforce'].

Similarly, we derived the popular domain and rare domain. Figure 10.5 shows the percentage of developers for the various domains.

Popular domain includes ['Full-stack developer', 'Back-end developer'].

Rare domain includes ['Marketing or sales professional', 'Educator or academic researcher'].

Analysing the second dataset, we came across skills demanded by the job description. We classify those skills into two categories—popular skills and rare skills.

Figure 10.6 gives the following insights

Popular skills include ['agile', 'javascript', 'java', 'sql'].

Rare skills include ['ml', 'flask', 'coldfusion', 'Cordova', 'grails', 'Pentaho', 'xquery', 'ember.js', 'verilog', 'laravel', 'go', 'spss', 'typescript', 'express', 'yarn','sqr', 'clojure', 'assembly', 'dojo', 'xen', 'xamarin', 'ada', 'cloudera', 'ruby on rails', 'r', 'self', 'storm'].

Similarly, we derived the popular domain and rare domain.

Popular domain includes ['Data or business analyst', 'Network Engineer', 'Software Developer/Java Developer'].

Rare domain includes ['Game or graphics developer', 'Back-end developer', 'Database Administrator(DBA)', 'Cloud Computing'].

We found a few insights from the data derived (see figure 10.7):

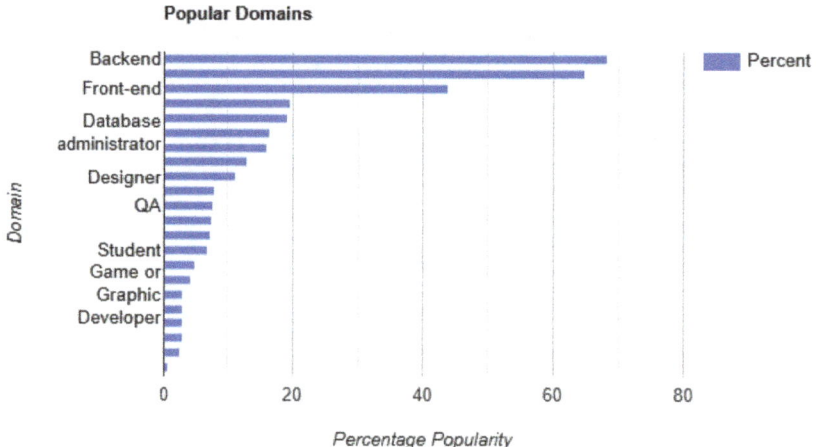

Figure 10.5. Bar graph of percentage domain present in the resume dataset.

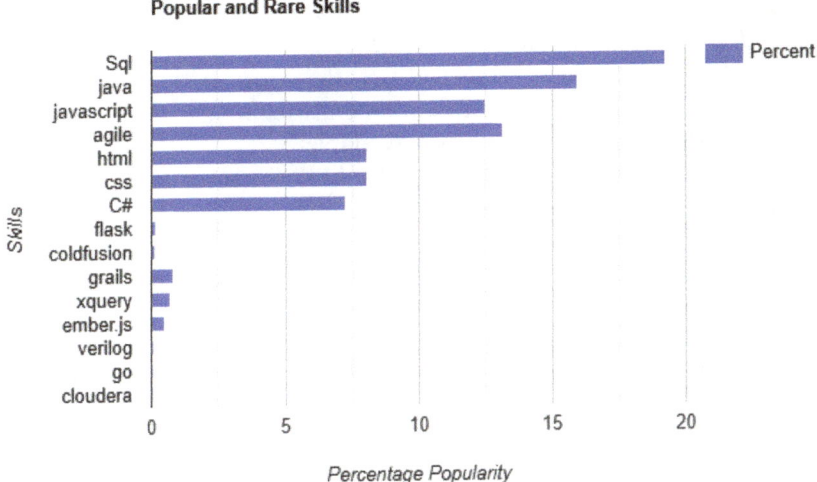

Figure 10.6. Bar graph of percentage skills present in the job description dataset.

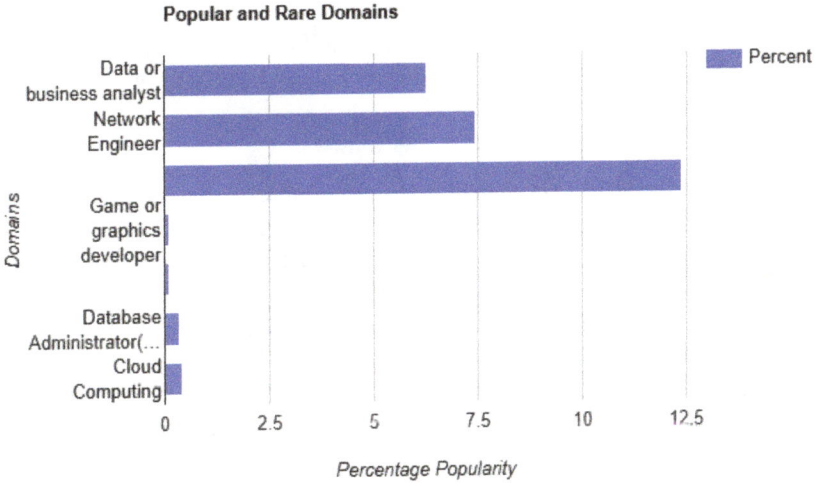

Figure 10.7. Bar graph of percentage domain present in the job description dataset.

- ['Clojure'] appears as a rare skill in both resumes and job descriptions. This suggests that Clojure is a skill that is relatively rare in both resumes and job descriptions, which could potentially make it a valuable skill to have if you are looking to stand out in the job market.

- Skills on the list of rare skills in job descriptions are generally more technical, while the skills on the list of rare skills in resumes are more diverse and include a mix of technical and non-technical skills. This suggests that the skills that are considered rare in job descriptions may be more in demand by employers looking for specific technical expertise, while the skills that are considered rare in resumes may be more valuable for demonstrating a broader range of abilities and experiences.

- According to resumes ['Backend'] is the most popular but in the job descriptions ['Backend'] is a rare domain. Meaning we have more backend developers than the opportunities available.
- ['Javascript', 'sql'] are popular skills for both the resume and job description datasets.

Q2. What is the distribution of resumes gender-wise corresponding to different ages?

Motivation: To find the impact of gender on job searches for different age groups.

Results and inference: Six groups are considered for analysing the resume gender-wise. Female resumes are skewed in comparison to male resumes. We found that more female resumes are in the age group 25–34 in comparison to the rest of the age groups. Below the age of 18 and after the age of 55, females are not interested in any kind of job. Table 10.1 shows the gender statistics corresponding to different age groups.

Q3. What is the distribution of salaries corresponding to different age groups?

Motivation: To find salary variation in different age groups.

Results and inferences: Salaries are given in dollars.

The most common salary groups for different ages found are given in table 10.2.

A few insights that can be drawn are as follows:

1. The most common salary group tends to increase as people get older. For example, the most common salary group for people under 18 years old is $500–$1500, while the most common salary group for people aged 45–54 is $5001–$10 000. This suggests that, on average, people tend to earn more authenticity, and that their salary may increase as they gain more experience and skills over time.

2. The salary groups for some age groups are more diverse than for others. For example, the salary group for people aged 25–34 includes both $1501–$5000 and $5001–$10 000, while the salary group for people aged 18–24 and 65 or

Table 10.1. Gender statistics corresponding to different age groups.

Gender Age	Female	Male	Other
18–24 years old	5.246 818	92.890 407	1.862 776
25–34 years old	6.559 225	91.975 656	1.465 119
35–44 years old	4.221 721	94.424 750	1.353 529
45–54 years old	4.191 617	95.059 880	0.748 503
55–64 years old	3.597 122	94.964 029	1.438 849
65 years or older	0.000 000	100.000 000	0.000 000
Under 18 years old	0.000 000	100.000 000	0.000 000

Table 10.2. Salary statistics corresponding to different age groups.

Age groups	Salaries
Under 18 years old	$500–$1500
18–24 years	$1501–$5000
25–34 years	$1501–$5000 and $5001–$10 000
35–44 years	$5001–$10 000
45–54 years	$5001–$10 000
55–64 years	$5001–$10 000 and $10 001–$16 000
65 years or older	$5001–$10 000

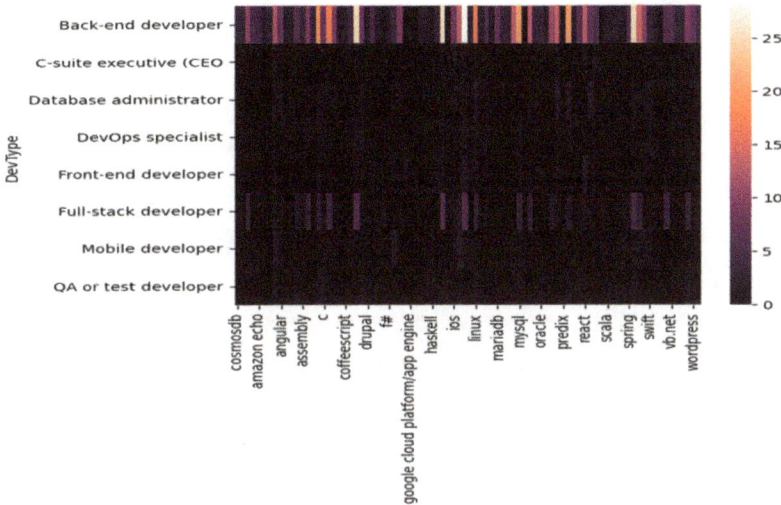

Figure 10.8. Heat map to correlate domain with skills.

older only includes $5001–$10 000. This suggests that the salary range for people in these age groups may be wider, potentially due to factors such as differences in experience, skills, or job roles.

Q4. What is the correlation between domain and skills?
Motivation: To find whether people have domain-centric or diverse skills.
Results and inferences: A heat map has been plotted to find the correlation between domain and skills.
We found a few insights from the map (in figure 10.8):
- Most people are backend developers.
- Full-stack developers and front-end developers occur frequently.
- 'Coffeescript' only occurs in the 'Mobile Developer' domain.
- 'Drupal' only occurs in 'Full Stack'.
- 'MariaDB' only occurs in 'Database Administrator' and 'Backend Developer'.

- 'CosmosDb' and 'Amazon Echo' are skills of a 'Backend Developer'.
- The most important skill for 'Backend Developer' is 'Haskell'.

Q5. What are the top skills used in various domains in resumes?

Motivation: To check if the most popular skills in the overall dataset prefer any domain.

Results and inferences: We used a bar graph to visualize the top three skills used by the developer in each domain and note that certain skills that are considered rare in one field may be very common in another.

We found a few insights from the graph (in figure 10.9):

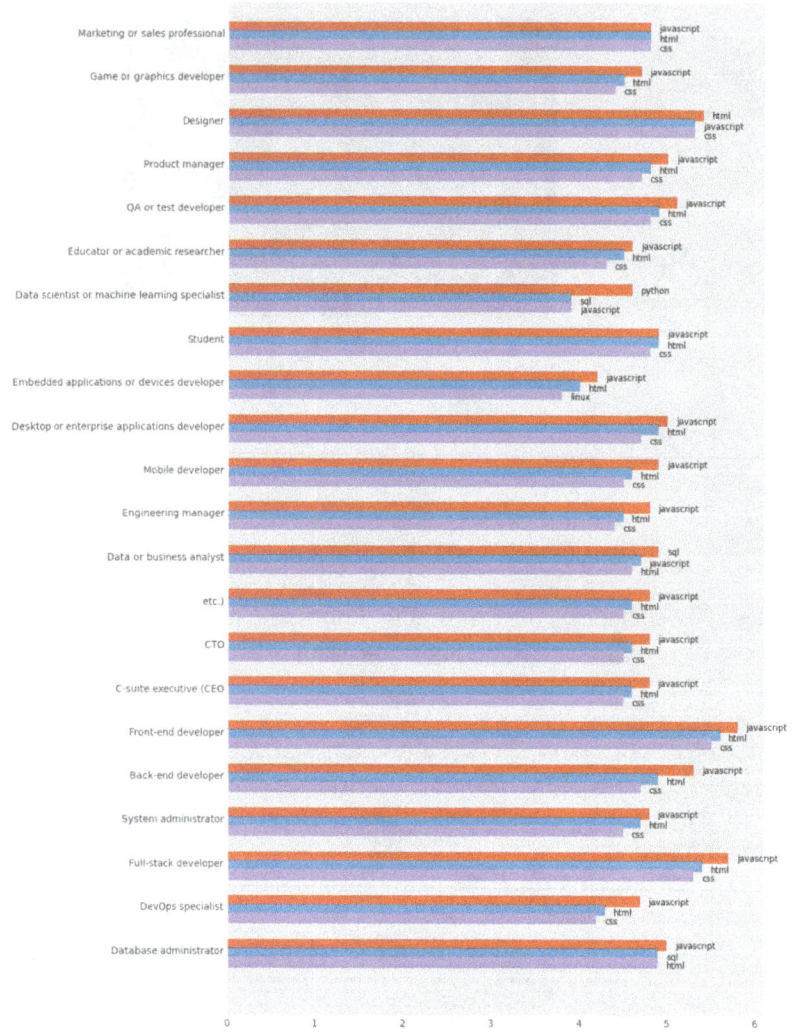

Figure 10.9. Bar graph of the top three skills worked with particular domain.

- Mostly the top three skills of every domain are ['javascript', 'html', 'css'].
- Domain types such as 'Data Administrator' and 'Data and Business Analyst' have top skills such as ['javascript', 'html', 'sql'].
- 'Embedded applications or device developer' is the only domain that consist of ['**Linux**'] as one of the top skills
- The 'Data scientist or machine learning specialist' domain has the most unique result where the top skills are ['**Python**', 'sql', 'javascript'].
- ['Python', 'Linux'] are not popular skills according to the result derived for the overall popular skills in the resume data, but here in a few cases, are popular skills.

In the second stage, we constructed a knowledge graph between the resumes and jobs available. And recommendations are done based on the number of connections formed between the resume node and the job description node. We form the connections with the skill and domain nodes in between. To rank our results we calculated the score based on 'commonSkillsCount' and 'commonDomainCount'.

We visualized resume data and job description data using a bipartite graph. Both of these datasets individually create their graphs and are joined using common skills and domains only. We can answer certain questions using only resume data, job description data, or both. As there is no direct connection between the resume and job description nodes, we used our recommendation algorithm to match job seekers with suitable jobs based on their skills and domains. The queries can be used to find jobs that match specific skill sets, identify the skills and experience of potential candidates, and calculate the compatibility between job descriptions and resumes. This information can be useful for job seekers and employers alike in finding and filling job opportunities.

Q6. Find all the job descriptions with similar skills as the resume.

Motivation: This query is useful for finding all the job descriptions that have similar skill requirements to a specific resume. This can be useful for job seekers who want to see which jobs might be a good fit for their skills and experience, and for employers who want to identify potential candidates with the skills they are looking for.

Results and inferences: In this step, we take all the potential JDs for a particular resume that may or may not be the best suitable recommendation. We show all the results even if there is only one connecting link between the JobID and the ResumeID (figure 10.10 shows the output for this).

Common neighbor score (link prediction)

This query is useful for calculating a score (in figure 10.11(b)) that represents the similarity between resumes and job descriptions (with KG in figure 10.11(a)). This can be useful for job seekers who want to see how similar their skills are to the jobs in the market.

"jobDes1"	"resume1"
{"name":"8aec88cba08d53da65ab99cf20f6f9d9"}	{"name":3}
{"name":"a9ced403587b3bef3a0a20f46be23a54"}	{"name":3}
{"name":"359170ce45b8499420a871bc717bc388"}	{"name":3}
{"name":"8c87853f2cb2977e05df98497bcb7583"}	{"name":3}
{"name":"c43743b1fcf4ac35abe0832bf4cf1ef7"}	{"name":3}
{"name":"af7cee118bb4f7dbdcd7c44f8515dc42"}	{"name":3}
{"name":"981d868a1f6136fcc517a90bf5001ff8"}	{"name":3}
{"name":"c353b8b1bb3e98475262b566dc8e8790"}	{"name":3}
{"name":"6050a89bc1eda4de4c8f362569d9181b"}	{"name":3}
{"name":"6fbca09aae852c35214c982d102a95e5"}	{"name":3}
{"name":"637e50c7560f7edae8f02ba95066239b"}	{"name":3}

Figure 10.10. The list of jobs that have similar skills to a particular resume. (Screenshot taken from Neo4j®.)

Adamic-Adar score

This query is useful for calculating the Adamic-Adar score (in figure 10.12(a)) for two nodes in the knowledge graph. This score represents the number of common neighbors that two nodes have divided by the log of the degree of each of those common neighbors. This can be useful for identifying nodes closely related in the knowledge graph (figure 10.12(b)) and recommending job seekers to jobs based on their common skills and domains.

The score for JD and resume

This query is useful for calculating a score (figure 10.13(a)) that represents the compatibility between a specific job description and a resume (with KG in figure 10.13(b)) and can be helpful for both job seekers and employers. Job seekers can use this information to understand how well their skills and experience align with the requirements of a specific job, while employers can use it to identify the most qualified candidates for a job based on their skills and experience.

Result verification using Adamic-Adar

This query is useful for verifying the results of the recommendation algorithm by calculating the Adamic-Adar score (in figure 10.14(a)) for each job description and resume pair (KG formed in figure 10.14(b)). By comparing the Adamic-Adar scores of different job descriptions and resumes, the algorithm can identify the most

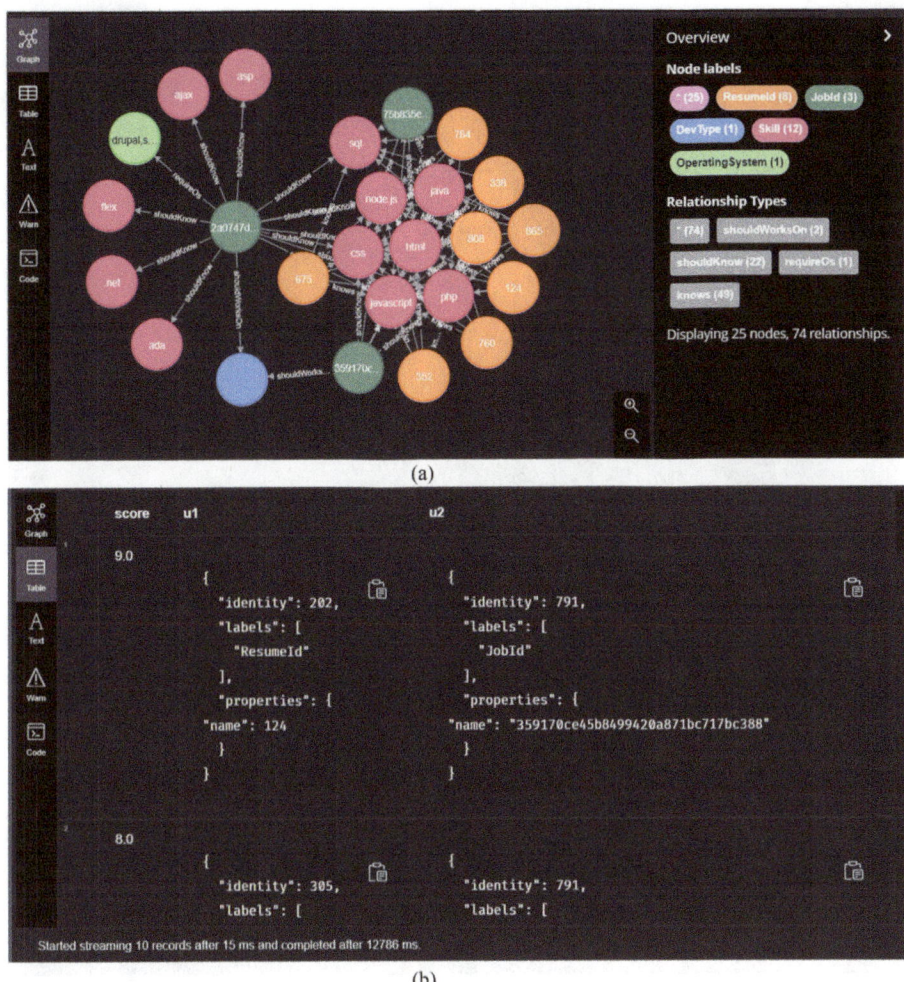

(a)

(b)

Figure 10.11. (a) Graph between a resume and different jobs. (b) Common neighbor score between a resume and different jobs. (Screenshots taken from Neo4j®.)

compatible matches and provide accurate and relevant recommendations to job seekers.

It is to be noted that in our work we cannot use the ROC curve as we do not have the ground truth regarding which resume took the job. We used two separate databases and joined them using various parameters (skills, domain), then recommended and ranked our matchings. We used the Adamic-Adar closeness score to give more reasons why our recommendation is accurate but we cannot verify it.

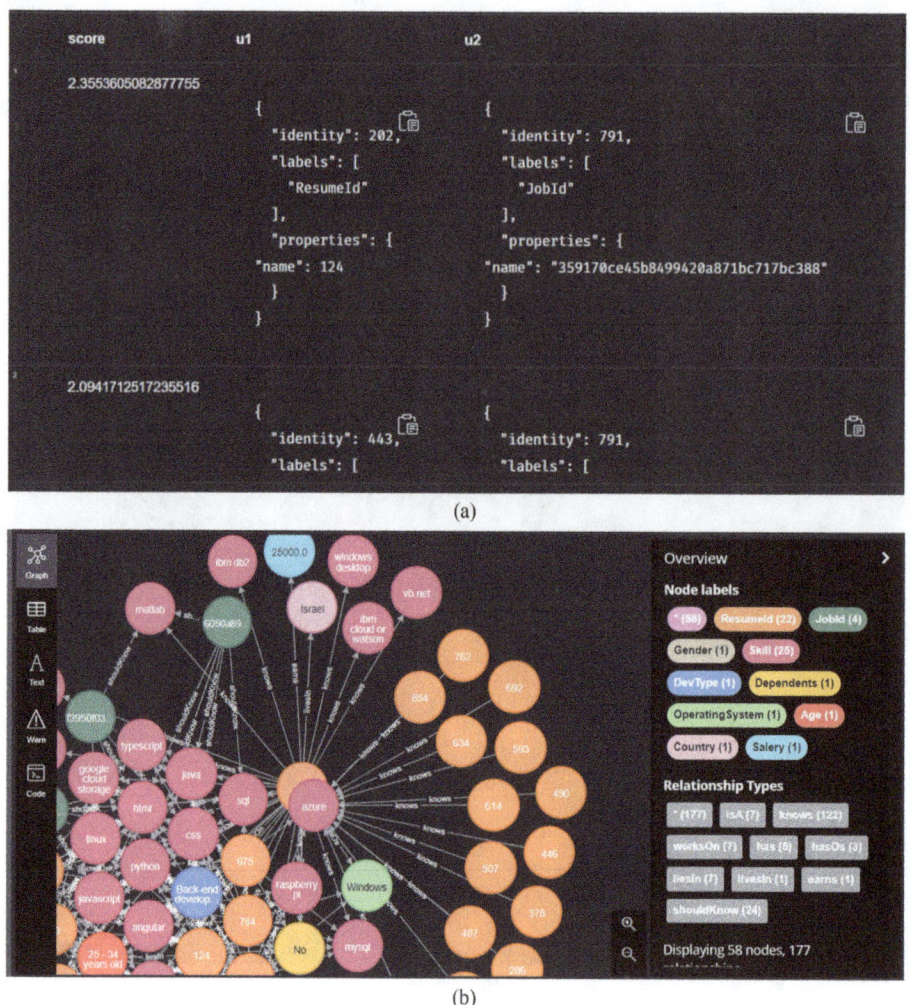

Figure 10.12. (a) Adamic-Adar score between a resume and different jobs. (b) Resume nodes are connected with job description nodes according to the Adamic-Adar score. (Screenshots taken from Neo4j®.)

10.5 Conclusion

Recommender systems have become very popular in online contexts due to their ability to generate both economic and social value. Online recruitment services, for example, use these systems to offer millions of job seekers tailored job openings. In this research, we developed and implemented a new recommendation system using a knowledge graph that can suggest jobs based on resumes and vice versa, to bring the full potential of advanced recommender systems to the job search domain. This system is designed to be scalable and reliable.

Our research using knowledge graphs showed that common abilities listed in resumes include Node.js, MySql, and CSS, while common abilities required in job

"adamic_score"	"common_neighbor_score"	"Resime"	"Job_Description"
2.3553605082877755	9.0	{"name":124}	{"name":"359170ce45b8499420a871bc717c388"}
2.0941712517235516	8.0	{"name":760}	{"name":"359170ce45b8499420a871bc717c388"}
1.9654892630364955	8.0	{"name":352}	{"name":"359170ce45b8499420a871bc717c388"}
1.9529309039059308	8.0	{"name":675}	{"name":"359170ce45b8499420a871bc717c388"}
1.741736095240056	6.0	{"name":543}	{"name":"f3950f031d1e12aefb4eb962e6f3702c"}
1.7398438573108084	6.0	{"name":543}	{"name":"6050a89bc1eda4de4c8f362569d9181b"}
1.7362518385703993	7.0	{"name":784}	{"name":"359170ce45b8499420a871bc717c388"}
1.7015529594471577	7.0	{"name":808}	{"name":"75b835ecdf3ffc91f3a65969554d88b0"}
1.6205430152031444	6.0	{"name":710}	{"name":"359170ce45b8499420a871bc717c388"}

(a)

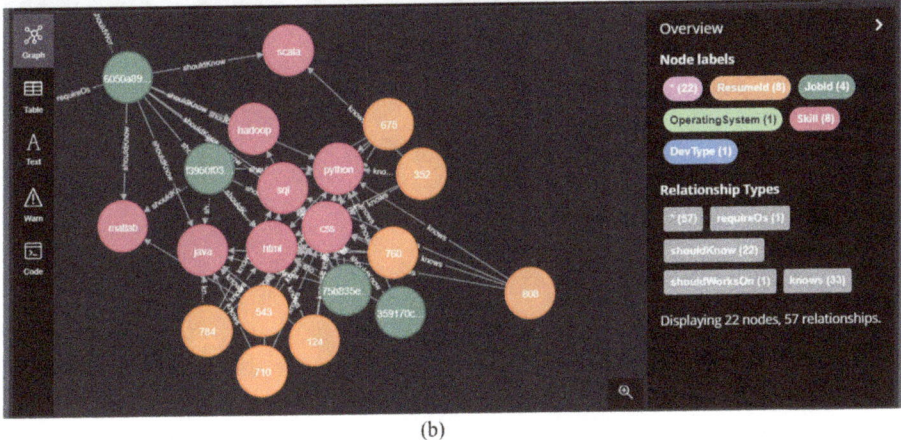

(b)

Figure 10.13. (a) Ordered list of common neighbor score and Adamic-Adar score between a resume and job. (b) Graphical representation of the resume nodes connected to job description nodes via skill nodes. (Screenshots taken from Neo4j®.)

listings include Javascript and Java. Our findings suggest that developers should focus on gaining proficiency in frameworks rather than just individual programming languages. We also found that developers tend to be more interested in full-stack and backend roles, while the job market currently needs more network engineers and cloud computing professionals. Our analysis of gender and age data showed that female resumes are most common in the 25–34 age group and that people tend to earn more as they gain experience and skills over time. Most people in our dataset were backend developers, and the most important skill for a backend developer was Haskell. We used common neighbor and Adamic-Adar methods to calculate scores between job descriptions and resumes and ranked the results according to these scores. Our Adamic-Adar verification indicates a strong relationship between recommended job descriptions and resumes.

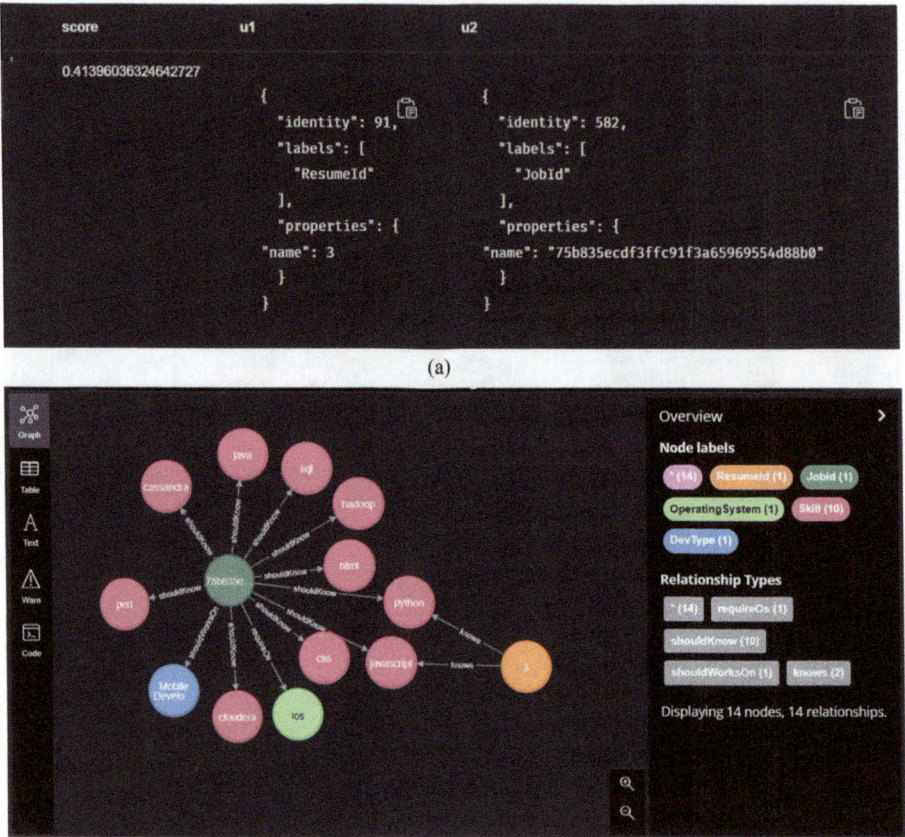

Figure 10.14. (a) Adamic score for the relationship between ResumeID 3 and JobID 75b83. (b) Graphical representation of the relationship between ResumeID 3 and JobID 75b83e. (Screenshots taken from Neo4j®.)

References

[1] Shury J, Vivian D, Spreadbury K, James A S and Tweddle M 2014 UK Commission's Employer Perspectives Survey 2014 https://assets.publishing.service.gov.uk

[2] Seaman A 2021 Post-pandemic job search trends LinkedIn (https://linkedin.com/news/story/post-pandemic-job-search-trends-4400521/)

[3] Zhang Y, Yang C and Niu Z 2014 A research of job recommendation system based on collaborative filtering *2014 Seventh IEEE Int. Symp. on Computational Intelligence and Design (Hangzhou, China)* pp. 533–8

[4] Guo X, Jerbi H and O'Mahony M P An analysis framework for a content-based job recommendation. *22nd Int. Conf. on Case-Based Reasoning (ICCBR) (Cork, Ireland)*

[5] Shalaby W, AlAila B, Korayem M, Pournajaf L, AlJadda K, Quinn S and Zadrozny W 2017 Help me find a job: A graph-based approach for job recommendation at scale *2017 IEEE International Conference on Big Data (Big Data)* pp. 1544–53

[6] Grida M, Fayed L and Hassan M 2020 A structured framework for building recommender system *J. Theor. Appl. Inform. Technol.* **98** 1101–14

[7] Bellini V, Biancofiore G M, Di Noia T, Sciascio E D, Narducci F and Pomo C 2020 GUapp: A Conversational Agent for Job Recommendation for the Italian Public Administration *2020 IEEE Conf. on Evolving and Adaptive Intelligent Systems (EAIS)* pp. 1–7

[8] Stack Overflow 2018 Developer Survey (https://kaggle.com/datasets/stackoverflow/stack-overflow-2018-developer-survey)

[9] US jobs on Dice.com (https://data.world/promptcloud/us-jobs-on-dice-com)

[10] Chicaiza J and Valdiviezo-Diaz P 2021 A comprehensive survey of knowledge graph-based recommender systems: technologies, development, and contributions *Information* **12** 232

[11] Ehrlinger L and Wöß W 2016 Towards a Definition of Knowledge Graphs 12th *Int. Conf. on Semantic Systems - SEMANTiCS 2016* **1695**

[12] Li N, Suri N, Gao Z, Xia T, Börner K and Liu X 2017 Enter a job, get course recommendations *Proc. iConference* vol 2

[13] Liu J and Duan L A 2022 Survey on Knowledge Graph-Based Recommender Systems *IEEE Transactions on Knowledge & Data Engineering* **34** 3549–68

[14] Hogan A, Blomqvist E, Cochez M, d'Amato C, Melo G D, Gutierrez C and Zimmermann A 2021 Knowledge graphs *ACM Comput. Surv. (CSUR)* **54** 1–37

[15] He M, Zhu Y, Lv N and He R 2022 A Feature Fusion-based Representation Learning Model for Job Recommendation *2022 2nd Int. Conf. on Consumer Electronics and Computer Engineering (ICCECE)*, Guangzhou, China, 2022, pp. 791–4

IOP Publishing

Data Analytics for Intelligent Systems
Techniques and solutions
Sachin Taran, Chhavi Dhiman and Manjeet Kumar

Chapter 11

A recommender system based on variants of singular value decomposition

Abhuday Tripathi, Ruchi Jain and Komal Tahiliani

The usage of recommender systems in e-commerce is evolving from being a niche innovation by a few web stores to a real commercial tool that is changing the face of the industry. Recommender systems (RSs) are a common tool to implement knowledge discovery in database (KDD) processes by large e-commerce companies to help clients. The primary components of any recommender system are a retrieval system, database system, ranking system, and machine learning models. It can be implemented in four ways: content-based filtering, collaborative filtering (CF), demographic-based recommender systems, and knowledge-based recommender systems. While content-based filtering uses the product's features and makes recommendations keeping the user's needs in mind, collaborative filtering based recommendations make predictions based on the past ratings of comparable users. Matrix factorization techniques can be used for implementing collaborating filtering, which would establish a relationship between user preference and interaction to generate recommendations. This chapter analyses the recommender system implementation using the singular value decomposition (SVD) algorithm. This is a collaborative filtering technique where we try to reduce the dimension of the input matrix. This results in a reduction in the dimension of the rating matrix. The SVD algorithm has two variants. Truncated SVD reduces the number of dimensions to a predefined value, while randomized SVD uses randomization techniques to compute factorization more effectively. We have computed the root mean square error (RMSE) and Spearman's rank correlation coefficient to measure the performance.

11.1 Introduction

The recommender system predicts a score to indicate a user's liking of any product. It can be expressed using a mathematical function f as given in

$$f: \text{User} \times \text{Products} \rightarrow \text{Rating Score}. \tag{11.1}$$

Since the introduction of laptops and smartphones, retail markets have been reorganized into companies focused on online commerce. Users now have access to a wider range of products to buy, so the amount of data that can be gathered has multiplied. There is a problem with information overload due to the World Wide Web's phenomenal growth. Users find it challenging to find what they need rapidly from the vast amounts of products available on a website.

In recent years, each customer can actively offer feedback based on customer involvement in social surveys conducted by e-commerce websites. This feedback is the ratings of products used by customers. e-commerce markets must now successfully utilize these data by developing new marketing strategies.

Additionally, e-commerce markets have actively implemented computerized personalization services to evaluate customer behaviour and purchasing habits. To suggest new, pertinent products to customers, e-commerce companies work to compile information about a variety of user interests. The collected data can be analysed in multiple ways depending on its type. In this respect, recommender systems have developed in the highly participatory environment of the Web. They supply lists of the products on the website that users will find very interesting and useful. It can work on databases of books, music, videos etc. A collaborative filtering based recommendation system is more popular than others [1]. Other techniques are content-based filtering, knowledge-based filtering system, hybrid filtering, etc.

A specific knowledge discovery in databases (KDD) technique is used in recommender systems (RSs). KDD systems employ a variety of subtle data analysis approaches to accomplish two overt objectives: first, to save money by identifying potential for efficiency and, second, to increase revenue by identifying new channels for customers to purchase more goods. KDD requires understanding the domain, preprocessing and cleaning, choosing the right algorithm, applying the algorithm, and finally, analysing and using the result. It is the backbone of the IT decision support system [2].

Companies are utilizing KDD, for example, to find out which products sell at given times of the year so they can increase the inventory sales in their retail stores and save millions in dollars by keeping less popular products out of stock. Other businesses use KDD to identify products that clients will find interesting by offering them a promotional offer, saving hundreds of thousands of dollars annually on advertising and promotional campaigns. These requirements typically involve using KDD to develop a new model and having a data analyst apply this model to the database and generate recommendations.

The recommender system must observe consumer behaviour while they are on the e-commerce site, learn from it, create a behavioural model, and use it to suggest products to customers that match their likes. The importance of KDD systems in e-commerce is immediately understood through recommender systems. They assist customers in locating the goods they want to purchase on the e-commerce website.

The most powerful recommender system technology available today, collaborative filtering, is employed in many popular e-commerce websites, including Amazon.com. In this technique, a rating matrix is generated based on the ratings provided by the users on the items. In this matrix, some values are missing, as the user will not rate every item in the matrix. Two different tasks can be performed with this matrix. We can predict the likelihood of a user liking a particular item, and the second outcome that can be obtained is to recommend N, the most useful item for the user, based on the matrix [3].

A system called collaborative filtering (CF) suggests products based on similarities. Collaborative filtering comes in two flavours: collaborative filtering based on features of items and collaborative filtering based on user similarity.

Collaborative filtering based on the similarity between users depends on the assumption that users will probably favour what other users like, provided they are similar, which is decided based on some metrics. The algorithm, therefore, first attempts to identify the user's neighbours based on similarities between two users before combining the rating score of the neighbouring user.

Figure 11.1 shows the neighbourhood construction using the nearest neighbour technique in a basic two-dimensional space. You will see that each user's neighbourhood consists of the other users determined by the proximity metric to be most similar to him. Asymmetric neighbourhoods are not required. The ideal neighbourhood for each user exists. A weighted composite of the neighbours' opinions about a particular product can be created once a neighbourhood of users has been identified. Thus the similarity measurement technique is a very important parameter in user-based CF algorithms. As the number of users increases, this technique starts taking more time, slowing down the entire system. To overcome this problem, we can use item-based collaborating filtering.

The item-based collaborative filtering technique predicts items by questioning and finding similarities between the given items and other items that have already been liked and rated by the users. On the other hand, when it comes to using the user's rating score, the algorithm for collaborative filtering based on the item is much the same as that for collaborative filtering based on users. The current user has already rated items; hence this algorithm determines how similar items are to the current item under suggestion rather than looking at the closest neighbours. Based on these item similarities, it then incorporates the customer's past preferences [4]. Two popular similarity measures are used with item-based CF. They are Pearson correlation coefficients and cosine-based similarity. There are different variants of Pearson correlation coefficients, such as the constrained Pearson correlation coefficient, weighted Pearson correlation coefficient, sigmoid function-based Pearson correlation coefficient, etc. The adjusted cosine measure is defined considering user preferences excluded in the cosine-based similarity [5].

The collaborative filtering engine's architectural scheme is modular. The user uses a Web interface to pick which products to promote to the user. The Web server software sends a request to the RS server. The recommender system has a database that keeps user, item, and rating data. It may also store comment data about each product. Data preprocessing must be performed before it is sent to the recommendation algorithm. The recommender system creates neighbourhoods and offers recommendations using data from its database of product ratings. The user sees the

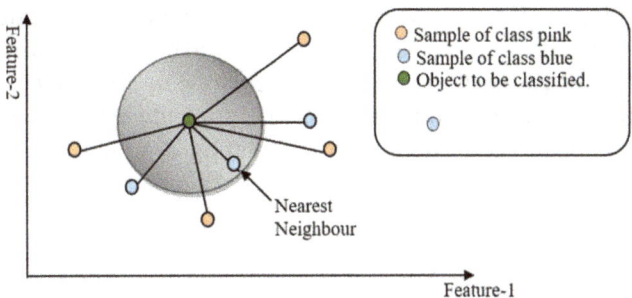

Figure 11.1. Formation of a neighbourhood.

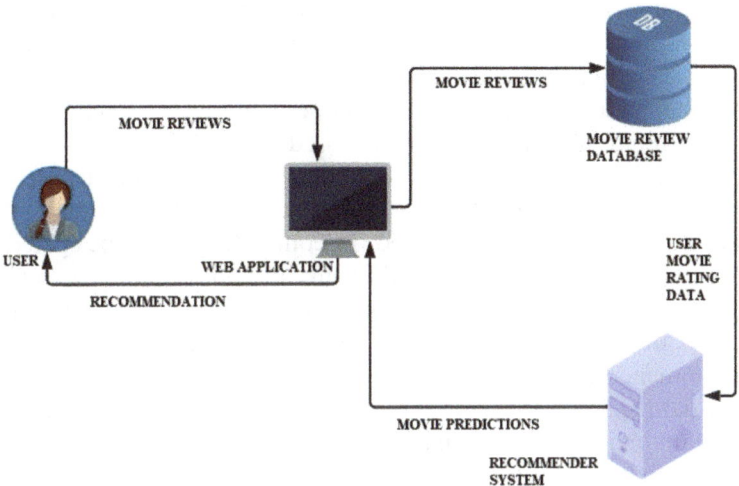

Figure 11.2. Architecture of the recommender system.

suggested products on the Web server software. This system should be robust and flexible, and it should be able to generate the result in a few milliseconds. It may also provide a ranking service where all the recommendations are ranked according to the user's need so that they can make intelligent decisions. As a recommendation is made based on a feature, its processing and storage must be handled in the recommender system. Overall the entire architecture can be divided into three main modules: the recommendation algorithm module, database, and user interface, as described in figure 11.2.

11.2 Limitations of the recommender system

The majority of collaborative filtering based RS creates communities of users with similar interests. As a proxy for proximity, the neighbourhood creation technique often uses distance measurement techniques such as cosine similarity, Pearson correlation, Euclidean distance measure, etc. These methods generate two different types of recommendations after determining the nearby neighbourhood.

1. *Estimation of the level of consumer C1 interest in product P1*: When using a correlation-based method, the forecast for interest of customer C1 for product P1 is calculated by first calculating a weighted sum of products that C1 and all of his neighbours have rated similarly and then by adding C's average rating to that.

2. *Top-N suggestions*: A consumer C1 is given a list of products that are recommended. The top-N suggestion is the term used to describe this. Once a neighbourhood has been established, the RS algorithm concentrates on the products the neighbours have rated and finds those N products that the consumer will find appealing.

It has been difficult to make accurate predictions and run recommendation systems as the number of users and items in every e-commerce website has continually grown along with the expansion of the Internet. Although these systems have been successful and are implemented properly in several domains, they appear to have some limitations [6], which are discussed below.

(A) *Data sparsity*: The system relies on direct feedback, such as user ratings of an item. Even though all users have participated in the rating process, the rating matrix might not be filled. In contrast, it might be sparse for every item offered. This will lead to difficulty in finding similar users [7].

Additionally, because consumers typically do not actively rank a linked group, it may be difficult to determine how comparable the items are. The recommendation system performs inaccurately as a result of these issues. The data sparsity is even the root cause of the cold-start issue.

Many pairings of customers do not correlate at all. They may not have any similarities as far as products are concerned. In reality, numerous commercial recommender systems assess enormous product collections. All the active customers may have given very few products in these systems a rating. As a result, Pearson's closest neighbour algorithms might be limited in how many products they can propose to a specific consumer. Reduced coverage is the issue, and this results from scarce neighbour ratings. Due to the limited amount of rating information that may be presented, suggestions may also be inaccurate.

(B) *Data scalability*: The system's response time should be kept constant while the size of the rating matrix increases. The data size may also increase in real time while the user interacts with the system. The algorithm should update and respond to the change. As the number of rating increases, the cost of computing the nearest neighbour and storage requirement also increases. Additionally, because the nearest neighbour algorithm took time to complete, RS could not reply to incoming requests rapidly and offer suggestions immediately. The scalability problem can also be reduced using genetic algorithm [8].

(C) *Name synonymy*: Different product names can refer to the same products in a real-world context. Correlation-based RS, which approaches these products differently, cannot find these latent associations. Take two clients as an illustration. One ranks ten different Lacto calamine goods as 'high', whereas the other also gives ten different calamine products a 'high' rating. The product sets in correlation-based RS would not match, making it impossible to find the hidden relationship that both users liked.

(D) *Dynamic nature of data*: As the data are generated online, everything about it changes dynamically, meaning neither its size is fixed nor its sparsity. The algorithm developed should be able to adjust to this, and it should be able to generate the result in the given time period.

11.3 Literature review

Most of the recommendation systems we use depend on the collaborative filtering algorithm to generate the recommendation based on linking other users to the product. Most RSs use the k-nearest neighbour algorithm or some variation to achieve this. The nearest neighbour algorithm discussed above has a huge computational cost as the distance has to be computed for a large matrix size. The matrix size increases continuously as the number of users rating the product also increases. Therefore we need to find ways to reduce the size of the matrix to reduce the computational cost. This would give a recommendation in real time. So we use singular value decomposition (SVD) to compute the factorization of the matrix such that the product of the factors gives an approximation of the matrix. It decomposes the matrix in three components, which are U, $\sum$, and V such that the actual matrix A can be written as

$$A = U \times \sum \times V^T. \tag{11.2}$$

The central matrix $\sum$ is a diagonal matrix with positive real numbers. The elements of this matrix are singular values of matrix A. This decomposition gives us the latent features of the rating matrix. As the matrix's size decreases, the computation matrix's overall cost also decreases. We can further decrease the time by using a truncated matrix. In this section, we will analyse the work done by other authors in the context of the recommender system.

In [9] the authors worked on improving the scalability of the SVD algorithm using an incremental approach. They developed the algorithm for online applications, so they generated the SVD factorization of only the incremented part of the online data and combined that with the factorization of the unchanged part. They experimented this algorithm, providing better accuracy and running time. In [10] the authors compared traditional user-based collaborative filtering techniques with SVD and the Pearson correlation coefficient. In [11] the authors used expectation maximization techniques with the SVD algorithm to implement a recommender system. This system provides better privacy to its user and can work in a distributed system. In [12] the authors used multi-objective immune and SVD algorithms to provide accuracy and diversity. In [13] the authors not only considered the rating provided by the user but their contextual information was also considered. This has made the system recommendation more personalized and increased the system's reliability. In [14] the authors used group-specific methods and SVD frameworks. This helped a lot in mitigating cold-start problems and improved prediction accuracy.

In [15] the authors used SVD and an item-based recommendation system to implement the RS. This reduced the existing problem of recommender systems such as scalability and sparsity problems. It has also improved the accuracy of the prediction. In [16] the authors used block-based matrix factorization and the SVD algorithm. They also used GPU processors and the CUDA framework. This algorithm shows improved performance in such environment. It has improved scalability in comparison to other algorithms. In [17] the authors implemented the SVD algorithm on a distributed system using Hadoop, Spark, and MPI technology. This has resulted in reduced time taken by the users. In [18] the authors tried to find a solution to the cold-start and sparsity problems. They used similarity measures between the contextual information, user, and item matrix. In [19] the authors used a social probability-based recommender system along with the SVD++ recommendation algorithm, resulting in improved performance.

Other factors can also be considered, such as attributes of the user and product, etc. In [20] the authors used a regularized SVD algorithm to implement a recommender system. It has a global optimal closed-form solution. This algorithm outperforms the standard SVD algorithm. In [21] the authors used an incremental algorithm to compute the SVD factorization of the matrix. This resulted in less time consumption and better accuracy. In [22] different categories of SVD algorithms were implemented by the authors. They were inspired by active learning techniques. The connection between the density of the matrix and prediction accuracy is also analysed. They provide better accuracy and efficiency than traditional matrix factorization-based RSs. In [23] effective neighbours for both the user and item are identified, and imputed data are computed. They are used with the SVD framework to implement a recommendation system, leading to better accuracy.

11.4 Singular value decomposition

An $m \times n$ matrix P is factored by the well-known matrix factorization method SVD into three matrices as given in [24]

$$P_{m\times n} = W_{m\times r} \times S_{r\times r} \times Z^T_{r\times n} \tag{11.3}$$

$$W \times W^T = I \tag{11.4}$$

$$Z \times Z^T = I. \tag{11.5}$$

Assuming r to be the rank of the matrix it is factored into three matrices W, S, and Z. Here W and Z are orthogonal matrices as described in equations (11.4) and (11.5), respectively. S is diagonal matrix of non-negative real numbers which can be written as $S = \mathrm{diag}(\sigma_1, \sigma_2, \sigma_3, \ldots, \sigma_k)$, where $\sigma_1 \geqslant \sigma_2 \geqslant \sigma_3 \geqslant \ldots \geqslant \sigma_k$ are positive real numbers.

In recommender systems we employ SVD to compute the hidden association between the user and items. After applying SVD we are left with a low-dimensional representation of the rating matrix. We apply the nearest neighbour algorithm on this matrix. Following that, we used it to create a list of the N products which are most relevant to the users.

In some recommender systems truncated SVD is used. It can be obtained from regular SVD by only storing the first k largest singular values and setting the rest of the positions to 0. Similarly we have to update the W and Z matrix. We delete the last $r-k$ columns of the matrix. Then the truncated matrix can be written as [25]

$$A_k = W_k \times S_k \times Z_k. \qquad (11.6)$$

This is the rank-k approximation of matrix A. We use truncated SVD to reduce the approximation but the accuracy of the result has to be taken care of also. The optimal value of k can be obtained. As SVD is computationally expensive we use truncated SVD in its place.

11.4.1 Generation of predictions

We start with a ratings matrix obtained from the feedback of customers. This matrix could be very sparse and is represented as S. We first need to eliminate the sparsity in the matrix. This can be done in many ways, such as computing the mean and replacing it with zero. This will not affect the latent features in the matrix. A sparsity problem arises as all users might not participate actively in the rating process, resulting in items being rated by very few users. This affects the outcome severely. It affects the similarity between user and user. Other techniques that can be used to reduce sparsity are dimensionality reduction of the rating matrix, using a bipartite graph of the user and items, and opting for item–item similarity are some of the methods that we can try [26].

We obtained a filled, normalized matrix R_n after normalization. R_n essentially equals $R + NP$, where NP is the matrix obtained after replacing the zero values. It is obtained using the above mentioned techniques to reduce sparsity. Applying the following methods, we apply matrix factorization of R_n to obtain an approximation of lower rank.

1. Factor R_n using the SVD algorithm to obtain the W, S, and Z^T matrices.
2. Reduce diagonal elements of matrix S to S_k which will have dimension k.
3. Calculate the square-root of S_k to obtain $S_k^{1/2}$.
4. Calculate the $W_k S_k^{1/2}$ and $S_k^{1/2} Z_k^T$ matrices.

The dot product of the ith row of $W_k S_k^{1/2}$ and jth column of $S_k^{1/2} Z_k^T$ can be used to compute the recommendation score between ith client and jth product [27].

11.4.2 Recommendation generation

The low-dimensional matrix thus obtained can be used to construct the cluster of a consumer and on the basis of their opinion, a recommendation of products can be made to a new consumer.

1. *Neighbourhood formation*: We can now form the cluster with the rating matrix having reduced dimension since its sparsity is less than the original rating matrix. In the same manner, we began with the initial rating matrix A and used the SVD algorithm to create the W, S, and Z matrices. Then we create the matrix S_k by decreasing the number of singular values in the diagonal matrix. We keep only the first k eigen values and and the remaining values are replaced by 0.

   In order to produce W_k and Z_k, we therefore performed dimensionality reduction. We calculates $W_k S_k^{1/2}$ using the same procedure as before. This $m \times k$ matrix represents the m consumers in k dimensions each of which represents some feature of the product. We can then build the cluster in that condensed space.

2. *Recommendation generation of top N*: Once the neighbourhood is established, we focus on the nearby customers and examine the goods they bought to suggest N goods that will be most relevant to the needs of the customer.

11.4.3 Performance evaluation

In reality, e-commerce websites receive a staggering number of daily client visits. The recommendation system used on such a website must be scalable enough to generate real-time product recommendations for this vast consumer base. The prediction generation method is typically split into two components by recommendation algorithms: offline and online components. The algorithm's offline component is the part that needs a lot of work, such as the computation of the SVD algorithm in a collaborative filtering based algorithm. The part of the algorithm known as the online component uses data from the offline component that have been stored to generate predictions for users dynamically.

11.5 Experiment

In this experiment, we have tried to evaluate the performance of the SVD algorithm. It can generate the ratings for every user, which are used to generate the recommendation. We have used the MovieLens database, which is made up of movie ratings generated over time and has been made available for public use. This dataset comes in multiple sizes and there are multiple variants of the dataset available [28]. In this dataset, ratings are available from different users for different kinds of movies.

The details of the number of users, movies, ratings, and size are provided in table 11.1. We have used all 100K, 1M, and 10M dataset. We wrote out the program in the Python language and calculated the results using the SCIPY library. The root mean square error (RMSE) of the given dataset is defined as the standard deviation of the prediction error. So the formula used to compute the RMSE is

$$\left[\frac{\sum_{i=1}^{N}\left(\text{Predict}_i - \text{Target}_i\right)^2}{N} \right]^{1/2}, \tag{11.7}$$

Table 11.1. Details of MovieLens dataset.

Data name	Number of users	Number of movies	Number of ratings	Size of the data
MovieLens 10M dataset	71 567	10 681	10 000 054	258 893 KB
MovieLens 1M dataset	6040	3952	1 000 209	24 018 KB
MovieLens 100K dataset	943	1682	100 000	1933 KB

Table 11.2. Comparison of SVD algorithm for different dataset.

MovieLens 100K dataset	RMSE	Spearman's rank correlation
Collaborative filtering	0.9181819871875179	0.9999999994941651
Singular value decomposition	7.980668035702924e−13	1.0
CUR	1.851604709434495	0.999999997942936

where N is the count of samples and (Predict$_i$–Target$_i$) is the ith difference as described in equation (11.7). Spearman's rank correlation coefficient indicates the relation between two variables. Its values are in the range $[-1,1]$. If it is 0 then it indicates there is no relationship between the two variable. A value very close to 1 or -1 indicates that one variable is a function of another variable. If the correlation coefficient is around 1 then both variables are associated in a positive way. In contrast, a value close to -1 indicates a negative relationship between both variables. The formula of the Spearman rank correlation coefficient is $r_s = 1 - \frac{6 \sum d_i^2}{t(t^2 - 1)}$. Here d_i is the difference in ranks of two observations and t is the count of observations. We compare the performance of SVD algorithm collaborative filtering and the CUR algorithm. We compute the RMSE and Spearman's rank correlation for the three different datasets described in the table. The results of the experiment are described in table 11.2.

From the table, we can easily conclude that SVD performs better than the collaborative filtering and CUR algorithms. The RMSE value and correlation coefficient are highly related. This is very much evident in our result for the SVD algorithm.

In the other experiment, we also applied truncated SVD on two datasets, i.e. the MovieLens 100K dataset and Movielens 1M dataset. In the truncated SVD algorithm, we try to find the given matrix's reduced rank decomposition by removing the diagonal matrix's singular values. We plotted the accuracy and the number of singular value graphs for the given dataset and computed the RMSE value for different component sizes used in truncated SVD. The component size ranges from 942 to 25, where the number of singular values obtained was 943.

We have randomly reduced the component value as the threshold data has to be identified. From the results, as given in figure 11.3, we can conclude that by reducing the number of components, we can reduce the memory requirement of the algorithm, but at the same time the RMSE value will also increase. Thus there is a limitation to this approach. Determining the threshold value for the component depends entirely on the problem we are solving and the dataset we are using. If accuracy is very important for our problem, we have to choose k as high as possible, and in the worst case, it will reduce to the standard SVD algorithm. If the computation cost has to be reduced, then we can reduce the number of components, but there has to be a pre-decided threshold value, as this will increase the RMSE error.

We have further tried to apply the same algorithm to the Movielens 1M dataset. This dataset has 6040 users and 3952 movies. Thus we have taken many singular values for our truncated SVD starting from 3950 to 50. We have demonstrated the result of the RMSE value for the different component sizes of the SVD algorithm. Our result demonstrates the exponential growth of the RMSE value as the number of components is decreased in figure 11.4.

The singular value decomposition algorithm becomes inefficient as the dimension of the matrix increases. In the recommendation system the size of the rating matrix increases continuously as new users are always coming along to give ratings. This makes the SVD algorithm very ineffective and sometimes it may fail altogether to give recommendations on time [29].

Many solutions have been provided which have led to multiple variants of the SVD algorithm. Some authors have tried to use random numbers to implement the SVD algorithm. This has led to a randomized SVD algorithm [30].

If A is a matrix whose factorization is required, then in a randomized SVD algorithm we try to find a random matrix Q such that the following equation is true:

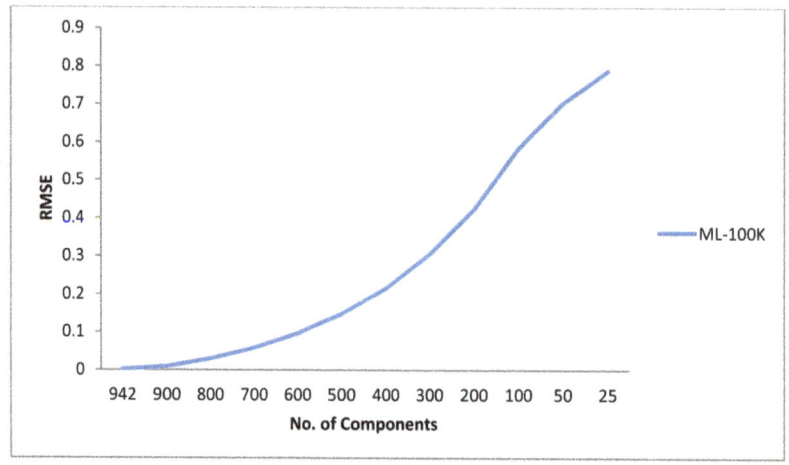

Figure 11.3. RMSE versus number of components for the ML-100K dataset.

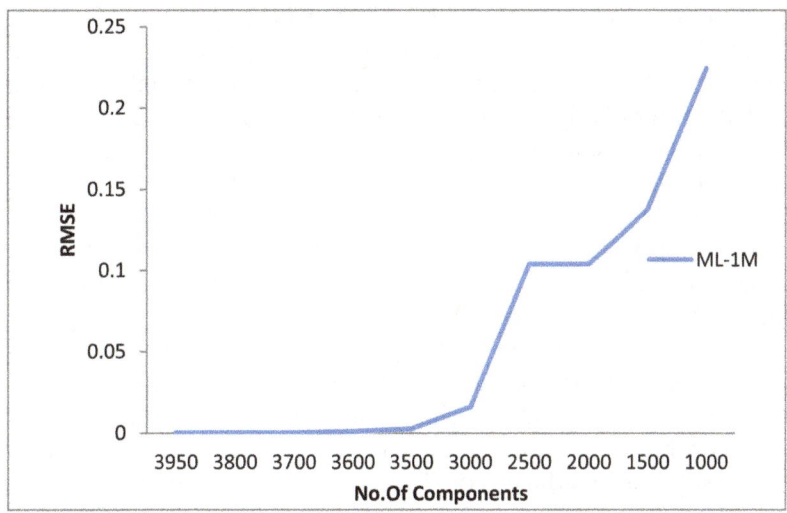

Figure 11.4. RMSE versus number of components for the ML-1M dataset.

Table 11.3. Performance of the randomized SVD.

Dataset	RMSE	Spearman's rank correlation
ML-1M	6.53E−15	1.0
ML-100K	3.28E−15	1.0

$$\|(I - QQ^T A)\| \leqslant \epsilon. \tag{11.8}$$

In this chapter we have also experimented with a randomized algorithm on our data and tried to evaluate its efficiency in terms of RMSE and Spearman's rank correlation.

We have applied the randomized SVD algorithm also on the given dataset which are ML-1M and ML-100K. The results are represented in table 11.3.

The results confirm that the randomized algorithm performs better than the SVD and truncated SVD algorithms on both datasets. The Spearman's rank correlation coefficient value is 1. It indicates perfectly correct factorization but a lower RMSE value indicates better performance of randomized SVD over truncated SVD.

11.6 Conclusion and future scope

The recommendation system is a very popular technique to increase sales on the e-commerce website. As the size of data is growing and diversity in data is growing, different techniques have been employed to compute recommendations for the user. Of the correlation-based methods, collaborative filtering, the SVD-based method, and the CUR algorithm are some techniques. In this chapter we showed that the SVD is the most promising of all of them, and it can be used to generate recommendations with the desired accuracy. We can use truncated SVD to reduce the computation cost if the accuracy is not required much. Therefore, we see that the online performance of SVD-based algorithms is significantly superior to correlation-based algorithms as we have reduced the dimension of the matrix. In low-dimensional space, neighbourhood formation is considerably quicker for the same reason. We also used two variants of the SVD algorithm: truncated SVD and randomized SVD. Of the two variants, randomized SVD performs better than truncated SVD. In the future we will try to improve the SVD algorithm more to obtain more scalability and work with a rating matrix that is changing in real time.

In this paper we analysed the SVD algorithm at a fundamental level. We have applied the three categories of the SVD algorithm to the MovieLens dataset. Although these are very good algorithms, in future, we would like to test other advanced algorithms on our dataset, such as higher-order SVD. The current algorithm considers a two-dimensional rating matrix between users and products. In real life, this matrix can have more dimensions, such as a rating matrix between the user, product, and time. We would also like to apply the tensor method to develop a recommender systems for such problems [31]. These matrices with higher dimensions can give more appropriate recommendations, but this comes at the cost of increased computation.

Acknowledgments

We would like to express our gratitude to Dr Keshavendra Choudhary, principal of SISTec and Professor Rahul Dubey, HOD of Computer Science and Engineering Department for all the guidance, support, and instructions they provided us throughout our journey. We would like to thank our colleagues for helping us directly and indirectly in completing this book chapter. The completion of this chapter would not have been possible without the guidance and support of our family and friends. Their belief in us has kept our spirits and motivation high during this process.

References

[1] Yang X, Guo Y, Liu Y and Steck H 2014 A survey of collaborative filtering based social recommender systems *Comput. Commun.* **41** 1–10
[2] Maimon O and Rokach L 2005 Introduction to knowledge discovery in databases *Data mining and Knowledge Discovery Handbook* (Boston, MA: Springer) pp 1–17
[3] Ekstrand M D, Riedl J T and Konstan J A 2011 Collaborative filtering recommender systems *Found. Trends® Hum.–Comput. Interact.* **4** 81–173
[4] Koohi H and Kiani K 2016 User based collaborative filtering using fuzzy C-means *Measurement* **91** 134–9
[5] Sarwar B, Karypis G, Konstan J and Riedl J 2001 Item-based collaborative filtering recommendation algorithms *Proc. of the 10th Int. Conf. on World Wide Web* pp 285–95
[6] Madhukar M 2014 Challenges and limitation in recommender systems *Int. J. Latest Trends Eng. Technol.* **4** 138–42

[7] Guo G 2013 Improving the performance of recommender systems by alleviating the data sparsity and cold start problems 23rd Int. Joint Conf. on Artificial Intelligence

[8] Georgiou O and Tsapatsoulis N 2010 Improving the scalability of recommender systems by clustering using genetic algorithms *Artificial Neural Networks—ICANN 2010* ed K Diamantaras, W Duch and L S Iliadis (Berlin: Springer) Lecture Notes in Computer Science vol 6352

[9] Zhou X, He J, Huang G and Zhang Y 2015 SVD-based incremental approaches for recommender systems *J. Comput. Syst. Sci.* **81** 717–33

[10] Chen V X and Tang T Y 2019 Incorporating singular value decomposition in user-based collaborative filtering technique for a movie recommendation system: a comparative study *Proc. of the 2019 the Int. Conf. on Pattern Recognition and Artificial Intelligence* pp 12–5

[11] Zhang S, Wang W, Ford J, Makedon F and Pearlman J 2005 Using singular value decomposition approximation for collaborative filtering *7th IEEE Int. Conf. on E-Commerce Technology* (Piscataway, NJ: IEEE) pp 257–64

[12] Chai Z Y, Li Y L, Han Y M and Zhu S F 2018 Recommendation system based on singular value decomposition and multi-objective immune optimization *IEEE Access* **7** 6060–71

[13] Gupta R, Jain A, Rana S and Singh S 2013 Contextual information based recommender system using singular value decomposition *Int. Conf. on Advances in Computing, Communications and Informatics* (Piscataway, NJ: IEEE) pp 2084–9

[14] Bi X, Qu A, Wang J and Shen X 2017 A group-specific recommender system *J. Am. Stat. Assoc.* **112** 1344–53

[15] Gong S, Ye H and Dai Y 2009 Combining singular value decomposition and item-based recommender in collaborative filtering *Int. Workshop on Knowledge Discovery and Data Mining* (Piscataway, NJ: IEEE) pp 769–72

[16] Bhavana P, Kumar V and Padmanabhan V 2019 Block based singular value decomposition approach to matrix factorization for recommender systems, arXiv:1907.07410

[17] Przystupa K, Beshley M, Hordiichuk-Bublivska O, Kyryk M, Beshley H, Pyrih J and Selech J 2021 Distributed singular value decomposition method for fast data processing in recommendation systems *Energies* **14** 2284

[18] Rodpysh K V, Mirabedini S J and Banirostam T 2021 Resolving cold start and sparse data challenge in recommender systems using multi-level singular value decomposition *Comput. Electr. Eng.* **94** 107361

[19] Kumar R, Verma B K and Rastogi S S 2014 Social popularity based SVD++ recommender system *Int. J. Comput. Appl.* **87** 33–7

[20] Zheng S, Ding C and Nie F 2018 Regularized singular value decomposition and application to recommender system, arXiv:1804.05090

[21] Sarwar B, Karypis G, Konstan J and Riedl J 2002 Incremental singular value decomposition algorithms for highly scalable recommender systems *5th Int. Conf. on Computer and Information Science* vol 1 pp 27–8

[22] Guan X, Li C T and Guan Y 2017 Matrix factorization with rating completion: an enhanced SVD model for collaborative filtering recommender systems *IEEE Access* **5** 27668–78

[23] Yuan X, Han L, Qian S, Xu G and Yan H 2019 Singular value decomposition based recommendation using imputed data *Knowl.-Based Syst.* **163** 485–94

[24] Chan T F 1982 An improved algorithm for computing the singular value decomposition *ACM Transac. Mathem. Soft. (TOMS)* **8** 72–83

[25] Zhou X, He J, Huang G and Zhang Y 2015 SVD-based incremental approaches for recommender systems *J. Comput. Syst. Sci.* **81** 717–33

[26] Papagelis M, Plexousakis D and Kutsuras T 2005 Alleviating the sparsity problem of collaborative filtering using trust inferences *Int. Conf. on Trust Management* (Berlin: Springer) pp 224–39

[27] Sarwar B, Karypis G, Konstan J and Riedl J 2000 Application of dimensionality reduction in recommender system—a case study *Technical Report* 00-043 Department of Computer Science, University of Minnesota, MN

[28] Maxwell Harper F and Konstan J A 2015 The MovieLens datasets: history and context *ACM Transactions on Interactive Intelligent Systems* **5** 19

[29] Drineas P, Kannan R and Mahoney M W 2006 Fast Monte Carlo algorithms for matrices II: computing a low-rank approximation to a matrix *SIAM J. Comput.* **36** 158–83

[30] Halko N, Martinsson P-G and Tropp J A 2011 Finding structure with randomness: probabilistic algorithms for constructing approximate matrix decompositions *SIAM Rev.* **53** 217–88

[31] Frolov E and Oseledets I 2017 Tensor methods and recommender systems *Wiley Interdiscip. Rev.: Data Min. Knowl. Discov.* **7** e1201

IOP Publishing

Data Analytics for Intelligent Systems
Techniques and solutions
Sachin Taran, Chhavi Dhiman and Manjeet Kumar

Chapter 12

Misleading multimodal news dataset for detecting fraudulent content

Deepika Varshney, Mohit Aggarwal and Neetu Saradana

With the rise in the popularity of social media, misinformation in the form of false or misleading content such as hoaxes, conspiracy theories, fabricated reports, click-bait headlines, and even satire is being widely disseminated with the goal of improving social media marketing effectiveness, boosting online traffic, increasing the number of followers for a page or business, creating a distraction, eliciting an emotional response, and shaping or changing public opinion. One of the greatest hurdles to their identification is the dearth of comprehensive datasets on misleading video and fake image detection that incorporate multiple categories. Hence, using a data extraction tool, a dataset consisting of forged images and misleading videos was created. The dataset consists of images, videos, and text from different social media platforms, thereby leading to multi-modality. The dataset created could now help in the identification of misleading information existing in any type of communication. This chapter presents the dataset along with the evaluation of the dataset to show where the dataset stands in terms of training and testing a model for the identification of fake news or misleading information. It also proposes the novelty of the dataset by testing variety of models and algorithms to predict misleading information. The algorithms used by different researchers and scholars have been taken into consideration and the results obtained are quite promising. Of all the models and algorithms applied, random forest and decision tree acquired the highest accuracy of 99.39%.

12.1 Introduction

'Fake news' can be defined as information that has no authentic evidence/sources or quotes perhaps, with the intention to propagate misleading news and lure users to view the news for some monetary incentives, such material can also be called 'click-bait'. With the advent of social media in recent years, fake news has proliferated, in part because it is so easily and quickly shared online. It is part of much larger ecosystem of misinformation and disinformation and encompasses much more than just false news stories. Some may contain a kernel of truth but lack contextualizing details, they may or may not contain any verifiable facts or sources. They may contain basic verifiable facts, but written in incendiary rhetoric, leaving out important details, or only presenting one point of view. This false information has adverse effects on society, for example, there was the tragic death of a 21-year-old student due to the propagation of false information via Baidu, China's largest search engine [1]. There are seven main distinct types of fraudulent content within our information ecosystem [2]: satire, misleading content, fabricated content, false context, manipulated content, hoaxes and clickbait headlines.

The difference between the existing dataset and proposed dataset is given in figure 12.1. One of the major roadblocks has been the lack of robust datasets for fake videos and manipulated images. The major drawback of the lack of model accuracy is the lack of availability of a dataset for the model to be trained and tested upon. Not only are available datasets scarce, but they also lack a plethora of features frequently required in research, such as news content and social context. Therefore, in this article, the fake news, manipulated images, misleading thumbnails, hoax news, click-bait and true news data repository Liar dataset is presented, which contains four comprehensive datasets with unique attributes such as news content, count, comment count, link to the post, date, and speaker/party.

The paper presents a misleading multimodal dataset that holds content from different social media platforms such as Instagram, YouTube, and Twitter (now knows as X). It consists of different attributes which are discussed and analysed

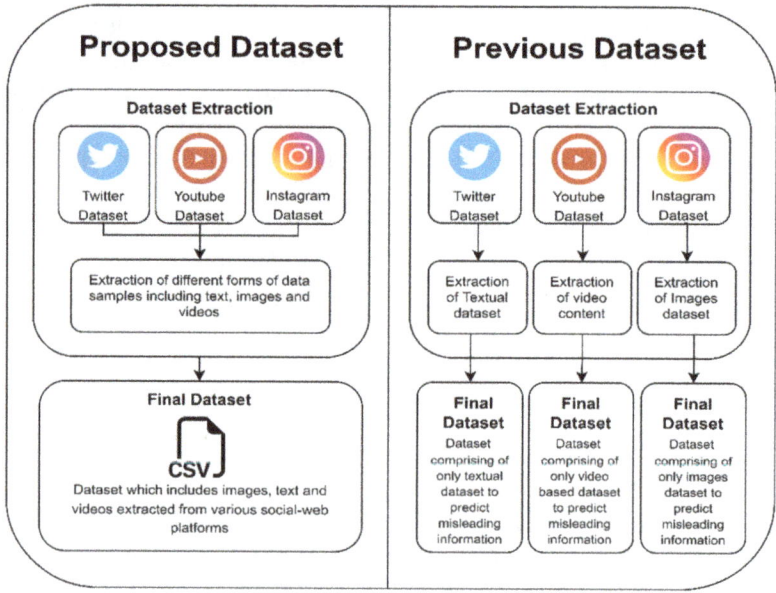

Figure 12.1. Difference between the existing dataset and proposed dataset.

based on their importance in the detection of fake news. The proposed dataset was prepared by extracting tweets, posts, and videos from different social media platforms such as Twitter, Instagram, and YouTube. This chapter also highlights the evaluation of the dataset by training and testing the proposed dataset on different models and algorithms for the detection of misleading information. We acquire the highest accuracy of 99.39% on implementing the random forest algorithm. From earlier research it has been identified that random forest works quite well for detecting fraudulent content [16, 17].

The following are the main contributions of this research work:

1. Proposing a multimodal dataset which incorporates the diverse forms of data present on social media platforms.
2. Proposing a dataset which comprises different forms of data such as text, images, and videos.
3. Testing the novelty of the proposed dataset by implementing different models and algorithms to predict fake news and misleading information.
4. Showing a comparative analysis of the existing fake news detection datasets with the proposed dataset.

The layout of the chapter is as follows. Section 12.2 comprises the proposed dataset, the extraction of the dataset, the description and analysis drawn from the dataset. Section 12.3 consists of the dataset experimentation, evaluation, and results, and focuses on evaluating the dataset by training and testing the dataset on different models and algorithms to detection fake news in any form, i.e. image, video, and text. The final section 12.4 comprises the conclusion.

12.2 Misleading multimodal dataset

In order to detect fake news, the most important factor is the availability of a variety of datasets on which various models and algorithms can be trained and tested. This section focuses on the proposed dataset for fake news and misleading information detection.

12.2.1 Dataset collection and annotation

To collect the most recent data from various platforms, the focus was primarily on three social media sites, Instagram, Twitter, and YouTube, between 2016 and 2020. There are numerous methods for obtaining data from social media websites. In this chapter Phatombuster.com, a cloud-based web scraping tool, was used to extract data from the social media sites [3]. To use this tool, one enters the URL of the account, and the tool will extract all of the information related to the account, including the date of posting, number of likes, comment count, and post description, and the result can be obtained in various formats including CSV and JSON, among others.

Table 12.1 represents the number of data entries extracted from three different social media platform for the preparation of the dataset.

Table 12.1. Number of data entries.

S. No.	Platform	Total number of entries
1.	YouTube	1568
2.	Instagram	1275
3.	Twitter	6630

After extraction of data from the three different social media platforms (YouTube, Instagram, and Twitter), the dataset was labelled for the identification of fake or misleading news. Two different annotators were used to manually annotate the entire dataset. The veracity was checked by the first annotator for the posts using fact checking websites including PolitiFact.com, FactCheck.org. This was done to label all three different extracted datasets for the detection of fake news and misleading information. The verification was checked by the second annotator to minimize the error in the dataset prepared. The following sections consist of the description of the datasets which were extracted from the three different social media platforms. It also covers the combining of the datasets and coming up with the final proposed dataset for the detection of fake news and misleading information.

12.2.2 Dataset description

Once the dataset was extracted from the three different platforms, the dataset was pre-processed to rule out the important attributes and make the dataset descriptive. As the dataset was extracted from multiple platforms, it includes all the types of communication that are possible on social media platforms. That is, all three categories, videos, images, and text, were included by extracting the dataset from YouTube, Instagram, and Twitter, respectively.

The dataset extracted from YouTube included a video link, video description, and other necessary attributes. Table 12.2 shows the final attributes of the YouTube dataset. Table 12.3 shows the final attributes of the Instagram dataset and table 12.4 shows the final attributes obtained from the Twitter dataset.

12.2.3 Proposed dataset analysis and evaluation

Dataset analysis refers to the identification of the essential points where the dataset stands out from the usual datasets that are being used. This section discusses the dataset analysis where we will cover the important features to understand the uniqueness of the proposed dataset. The graphs show that the proposed dataset comprises a majority of textual posts such as tweets, thoughts, and many more posted on different social media platforms, followed by images such as Instagram posts and Twitter posts, where users have shared different types of thoughts and opinions in the form of images such as opinion news headlines, updates, personal information, etc. The least posted source of communication is videos. Mostly the

Table 12.2. Final attributes after annotation of the YouTube dataset.

S. No.	Name of the attribute	Description of the attribute
1.	Link	The link of the video uploaded on YouTube.
2.	Speaker/party	The author of the video uploaded.
3.	Category	The genre of the video uploaded.
4.	Platform	The social media platform on which the video was uploaded.
5.	Like count	The number of likes on the video.
6.	Comment count	The number of comments on the video.
7.	Description	A caption/description for the video uploaded.
8.	Veracity	Annotation 1
9.	Verification	Annotation 2

Table 12.3. Final attributes obtained from the Instagram dataset.

S. No.	Name of the attribute	Description of the attribute
1.	Link	The link of the post uploaded on Instagram.
2.	Date	The date on which the post was posted.
3.	Speaker/party	The author of the post.
4.	Category	The genre of the post uploaded.
5.	Platform	The social media platform on which the post was uploaded.
6.	Like count	The number of likes on the post.
7.	Comment count	The number of comments on the post.
8.	Description	A caption/description for the post.
9.	Veracity	Annotator 1
10.	Verification	Annotator 2

Table 12.4. Final attributes obtained from the Twitter dataset.

S. No.	Name of the attribute	Description of the attribute
1.	Link	The link of the post uploaded on Twitter.
2.	Date	The date on which the tweet was posted.
3.	Speaker/party	The author of the tweet.
4.	Category	The genre of the tweet uploaded.
5.	Platform	The social media platform on which the tweet was uploaded.
6.	Like count	The number of likes on the tweet.
7.	Comment count	The number of comments on the tweet.
8.	Description	A caption/description for the tweet.
9.	Veracity	Annotator 1
10.	Verification	Annotator 2

videos are posted on YouTube, but sometimes people also share videos on Instagram in the form of reels or stories where people share their opinion over any situation, which can be political, entertainment, stress, etc. Another way of analysing the proposed dataset is based on the platform that has been used by the users to post their comments or viewpoints on any topic.

Figure 12.2 shows the division of the proposed dataset based on the social media on which the dataset was posted. Clearly, we can see from the graph that Twitter is the most widely used platform when it comes to expressing viewpoints on any type of situation. The next most common platform for sharing content is YouTube. Users generally share content by recording their viewpoint along with the evidence and additional resources they want to share with everyone to validate their viewpoint with confidence.

Table 12.5 is the tabular representation of the number of data samples marked as true and false for each of the three social media platforms for the formation of our proposed dataset.

Based on the proposed dataset, one of the most interesting applications of our dataset is that it will provide assistance in the development of various machine learning models and algorithms for the detection of fake or misleading information. The dataset is framed to identify misleading information from videos, images, and

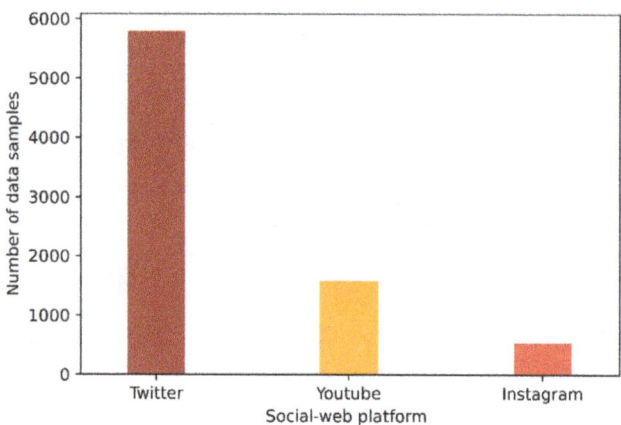

Figure 12.2. Platform division.

Table 12.5. The number of true and false labelled data samples.

Social media platform	Data samples labelled true	Data samples labelled false	Total data samples
YouTube	1000	586	1586
Instagram	533	6	539
Twitter	4060	1736	5796

text, which sums up every type of communication resource on social media [4–7]. Thus, the dataset will be able to identify and answer the following questions:

1. Given the dataset, how will the machine learning or deep learning algorithms be able to identity misleading information spread across different social media platforms?
2. Based on surface-level linguistic realizations only, how well can machine learning algorithms classify a short statement into a fine-grained category of fake news?
3. There exist different forms of data on social media handles, so on what dataset must be the algorithms be trained to obtain an overall model or working system for the identification of misleading information or fake news?

12.3 Experimental analysis and results

This section focuses on the experimental analysis findings of the proposed dataset. The proposed dataset can be evaluated by splitting the dataset into training and testing datasets for the prediction of fake or misleading information.

Thus, in order to split the dataset into testing and training, the data entries with descriptions are taken out. That is, 1993 entries which include data from YouTube, Instagram, Twitter, have a description along with the post. Out of 1993 data entries, 1494 were considered as the training dataset and 499 were taken as the test dataset for the detection of misleading information on different social media platforms.

Once the training and test datasets was split, different models and algorithms were used to evaluate the novelty of the proposed dataset. The results obtained from the models/algorithms are represented using various forms, one of which is the confusion matrix. The confusion matrix [14] is a pictorial representation in the form of a table that depicts the outcome of any model statistically. It consists four major values: the true-positive (TP) value, the true-negative (TN) value, the false-positive (FP) value, and the false-negative (FN) value. The TP and TN values are present diagonally to each other, while the FP and FN values are presented on the other diagonal, as shown in figure 12.3.

Therefore, from the above confusion matrix and the four types of values, the accuracy can be calculated by the equation

$$\text{Accuracy Score} = \frac{\text{Correct Predictions}}{\text{All Predictions}} = \frac{\text{TP} + \text{TN}}{\text{TP} + \text{FN}}. \tag{12.1}$$

Another method for representing the accuracy obtained by testing the proposed model to predict the labels of multilingual toxic comments is the receiver operating characteristics (ROC) curve. The ROC curve can be computed as the true-positive rate (TPR) and the specificity as the false-positive rate (FPR):

$$\text{TPR} = \frac{\text{True Positive}}{\text{All Actual Positive}} = \frac{\text{TP}}{\text{TP} + \text{TN}} \tag{12.2}$$

Actual Values

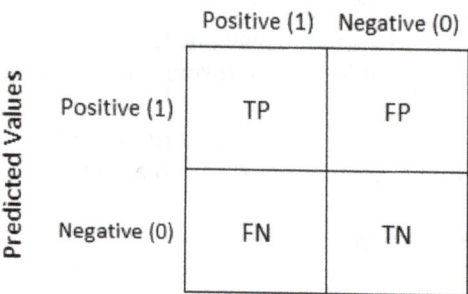

Figure 12.3. Description of a confusion matrix.

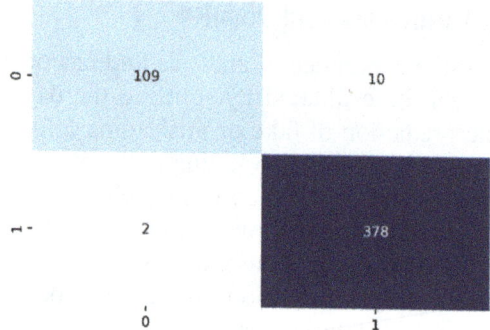

Figure 12.4. The confusion matrix of the SVM.

$$\text{FPR} = \frac{\text{False Positive}}{\text{All Actual Negative}} = \frac{\text{FP}}{\text{TN} + \text{FP}}. \qquad (12.3)$$

The area under the curve (AUC) [15] is the most crucial feature to analyse the ROC curve. It helps to analyse precision of the proposed model. For the AUC, the convention is that the more the curve shifts towards the top right corner, the more value of AUC increases, which results in increased accuracy of the proposed model.

Figures 12.4 and 12.5 show the confusion matrix and ROC curve obtained after testing the support vector machine (SVM). The curve shows some effective results, as shown in the figures.

Figures 12.6 and 12.7 represent the confusion matrix and ROC curve, respectively, after implementing the naïve Bayes algorithm. The confusion matrix represents the true and false number of predictions made by the naïve Bayes algorithm.

Next, figures 12.8 and 12.9 show the confusion matrix and ROC curve obtained by implementing the decision tree classifier. The prediction obtained shows promising results for the true-positive and true-negative results.

Figures 12.10 and 12.11 show the results obtained after implementing the random forest model/algorithm. The ROC curve obtained the highest AUC score as well as

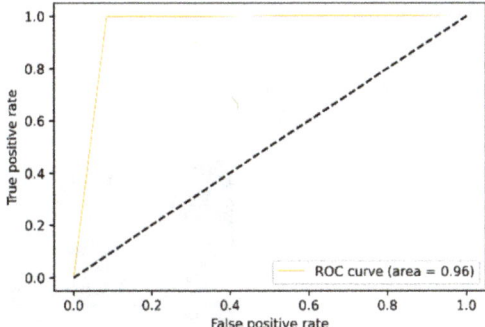

Figure 12.5. The ROC curve of the SVM.

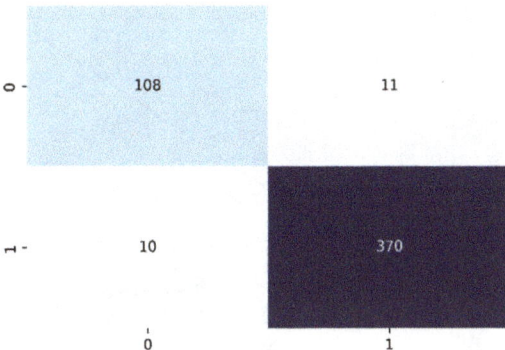

Figure 12.6. The confusion matrix for naïve Bayes.

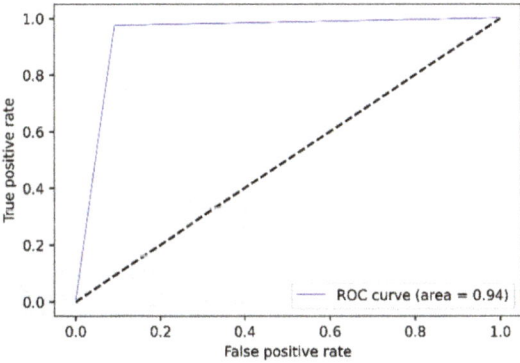

Figure 12.7. The ROC curve for naïve Bayes.

the true-positive and true-negative rates. Therefore, the results seem quite good and accurate for the detection of fake or misleading information.

Figures 12.12 and 12.13 present the performance measures of the ROC curve and confusion matrix after implementing gradient boosting. The model was split into training and test datasets which were then tested upon the gradient boosting

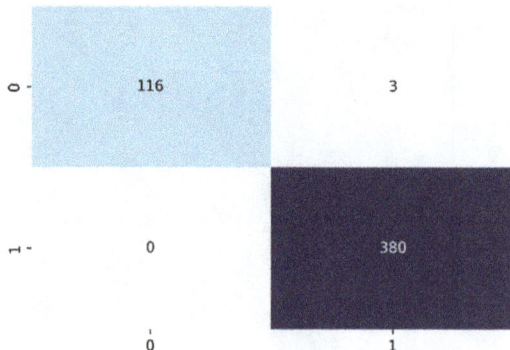

Figure 12.8. The confusion matrix for the decision tree.

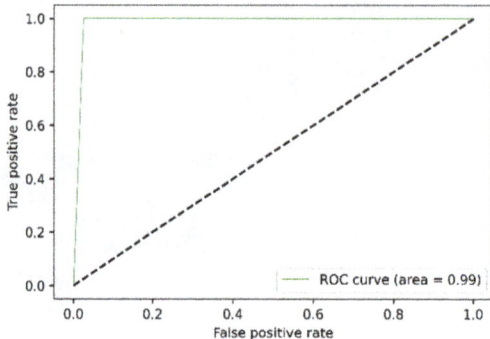

Figure 12.9. The ROC curve for the decision tree.

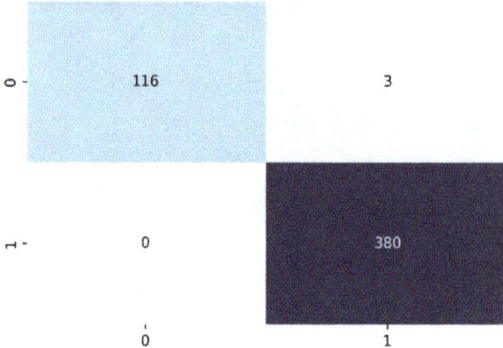

Figure 12.10. The confusion matrix for the random forest.

algorithm. The results obtained appear to support the accuracy of the proposed dataset.

The proposed dataset was trained on the recurrent neural network (RNN). Figures 12.14 and 12.15 show the training accuracy and training loss obtained upon training the RNN model with the training dataset.

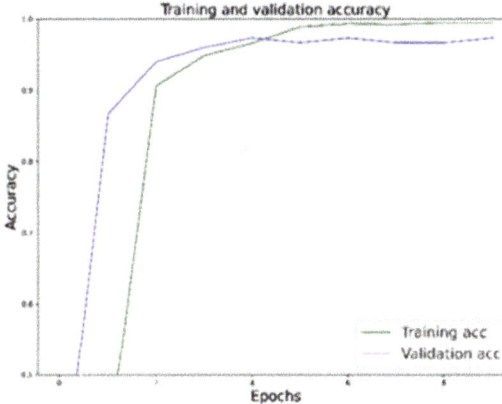

Figure 12.11. The ROC curve for the random forest.

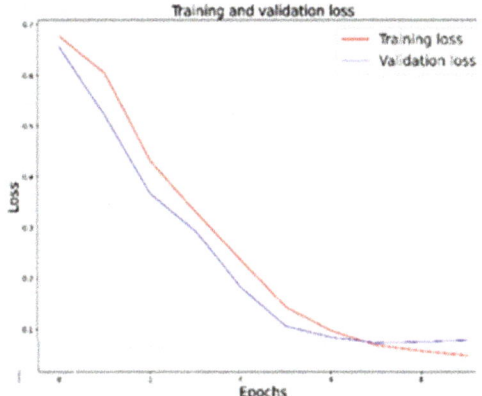

Figure 12.12. The confusion matrix for gradient boosting.

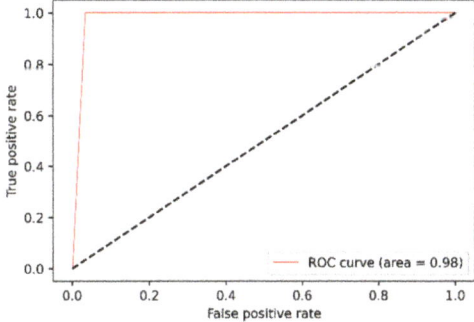

Figure 12.13. The ROC curve for gradient boosting.

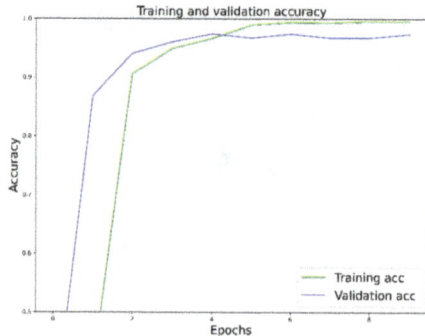

Figure 12.14. The training accuracy of the RNN.

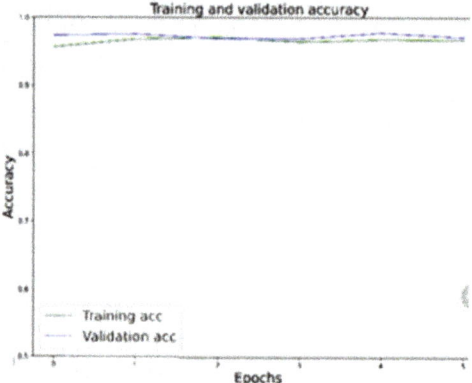

Figure 12.15. The training loss of the RNN.

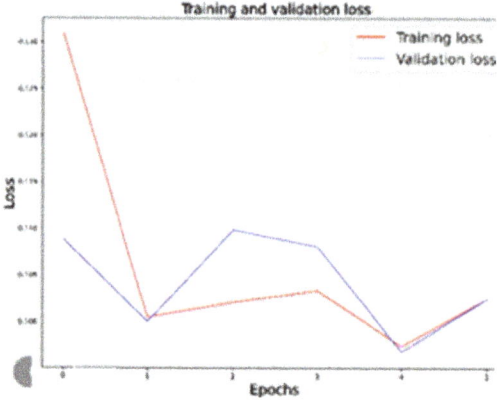

Figure 12.16. The confusion matrix for the RNN.

Figures 12.16 and 12.17 show the confusion matrix and the ROC curve obtained after testing the proposed dataset on the RNN. This model falls under the category of deep learning, which further shows the novelty of the dataset upon training it on a deeper level of models and algorithms for the detection of misleading information.

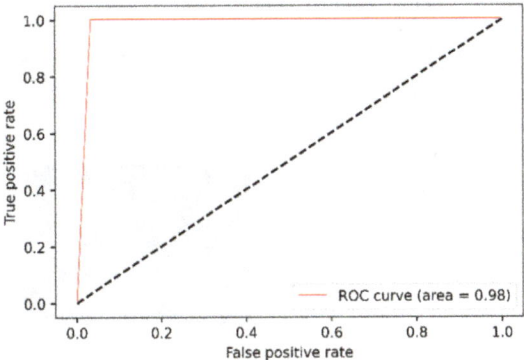

Figure 12.17. The ROC curve of the RNN.

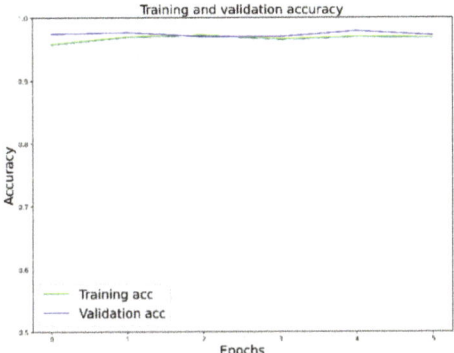

Figure 12.18. The training accuracy of the LSTM.

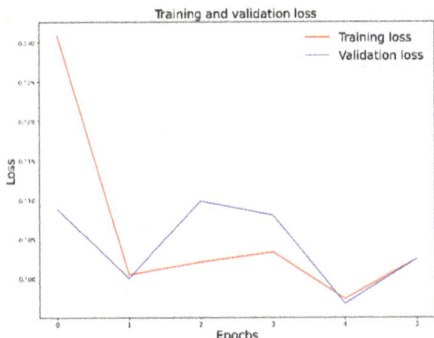

Figure 12.19. The training loss of the LSTM.

The proposed dataset was trained on the next model, LSTM. Figures 12.18 and 12.19 show the training accuracy and training loss obtained upon training the Long Short Term Memory (LSTM) model with the training dataset.

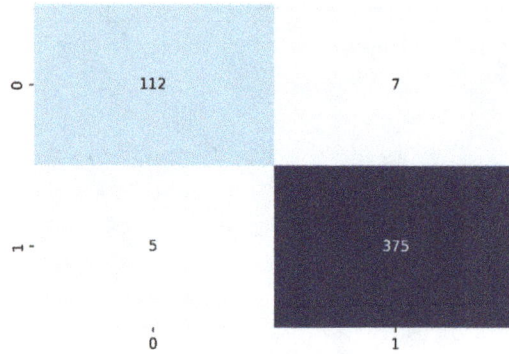

Figure 12.20. The confusion matrix of the LSTM.

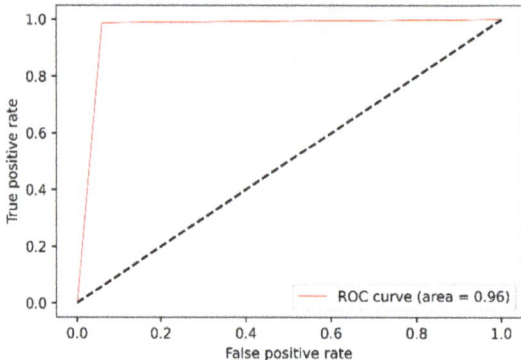

Figure 12.21. The ROC curve of the LSTM.

Further, figures 12.20 and 12.21 depict the confusion matrix and ROC curve obtained upon testing the proposed dataset on the deep learning-based model LSTM.

Table 12.6 presents a list of the results obtained after testing the models on the proposed dataset, which includes the testing matrix obtained for each of the models/algorithms implemented.

Thus, on obtaining the results after testing different models and algorithms on the proposed dataset, we can see from table 12.6 that all the models and algorithms obtained an accuracy above 95%. The highest accuracy was obtained by random forest, 99.39%, which seem to be quite promising.

Table 12.7 shows a comparative study of the results acquired upon training and testing our proposed misleading multimodal dataset, which shows a highest accuracy of 99.39%. This proves that the proposed dataset obtains good results and can be considered as a unique dataset with different types of communication resources such as images, videos, and text for the detection of fake or misleading information [18]. This proves the novelty of the proposed dataset in terms of accuracy, or prediction of misleading information or fake news.

Table 12.6. Comparative analysis of the results acquired.

S. No.	Model/algorithm	Class predicted	Precision	Recall	F1 score	AUC score	Accuracy
1.	SVM	FALSE	0.98	0.92	0.95	0.9553	0.9759
		TRUE	0.97	0.99	0.98		
2.	Naïve Bayes	FALSE	0.92	0.91	0.91	0.9406	0.9579
		TRUE	0.97	0.97	0.97		
3.	Decision tree	FALSE	1	0.97	0.99	0.9873	0.9939
		TRUE	0.99	1	1		
4.	Random forest	FALSE	1	0.97	0.99	0.9873	0.9939
		TRUE	0.99	1	1		
5.	Gradient boosting	FALSE	1	0.97	0.98	0.9831	0.9919
		TRUE	0.97	1	0.99		
6.	RNN	FALSE	0.98	0.96	0.97	0.9763	0.9859
		TRUE	0.99	0.99	0.99		
7.	LSTM	FALSE	0.96	0.94	0.95	0.9640	0.9759
		TRUE	0.98	0.99	0.99		

Table 12.7. Comparative study of the results acquired upon training and testing our proposed misleading multimodal dataset.

S. No.	Proposed model/ algorithm	Accuracy obtained using different datasets	Accuracy obtained upon testing our proposed dataset
1.	SVM [8]	0.9664	0.9759
2.	Naïve Bayes [9]	0.9200	0.9579
3.	Decision tree [10]	0.9500	0.9939
4.	Random forest [10]	0.9635	0.9939
5.	Gradient boosting [11]	0.8250	0.9919
6.	RNN [12]	0.9200	0.9859
7.	LSTM [13]	0.9108	0.9759

12.4 Conclusion

Misleading information is one of the growing concerns in today's digital era. With users being engaged more in online social media platforms, misleading information has found a new route for its wide spread. Thus, in order to identify misleading information on various multimedia platforms, this chapter presents a multimodal dataset. The dataset consists of various forms of data samples which include every form of data that can be shared or accessed on different online platforms. It is a multimodal dataset which is extracted and compiled from various social media platforms such as YouTube, Twitter, and Instagram. The proposed dataset includes different forms of data samples, i.e. images, text, and videos, which are the means of spreading misleading information across all the online platforms. This chapter also presents the novelty of the proposed dataset by implementing different models/ algorithms proposed by various researchers and scholars who have been able to predict misleading information on individual forms of dataset. The highest accuracy obtained was from random forest, which was 99.39%, which seems quite promising. Thus, this chapter proposes a new dataset that can be considered for the betterment of various studies to identify misleading information in any form of data, be it videos, text, or images. It can be enhanced further in order to identify the misleading or false information being spread through different social media platforms.

References

[1] Desai S, Mooney H and Oehrli J A 2021 'Fake news,' lies and propaganda: how to sort fact from fiction *University of Michigan Library* https://guides.lib.umich.edu/fakenews (Accessed: 29 December 2021)

[2] Wardle C 2017 Fake news. It's complicated 17 February *First Draft News* https://firstdraftnews.org/articles/fake-news-complicated/ (Accessed: 29 December 2021)

[3] https://phantombuster.com/

[4] Mena P, Barbe D and Chan-Olmsted S 2020 Misinformation on Instagram: the impact of trusted endorsements on message credibility *SAGE J.*

[5] Instagram 2019 Combatting misinformation on Instagram *Meta* 16 December https://about.instagram.com/blog/announcements/combatting-misinformation-on-instagram (Accessed: 18 January 2022)

[6] Boididou C, Papadopoulos S, Zampoglou M, Apostolidis L, Papadopoulou O and Kompatsiaris Y 2018 Detection and visualization of misleading content on Twitter *Int. J. Multimed. Inform. Retr.* **7** 71–86

[7] Zhang X and Ghorbani A A 2020 An overview of online fake news: characterization, detection, and discussion *Inf. Process. Manage.* **57** 102025

[8] Hussain M G, Hasan M R, Rahman M, Protim J and Hasan S A 2020 Detection of Bangla fake news using MNB and SVM classifier arXiv: 2005.14627

[9] Adiba F I, Islam T, Kaiser M S, Mahmud M and Rahman M A 2020 Effect of corpora on classification of fake news using naïve Bayes classifier *Int. J. Autom., Artif. Intell. Mach. Learn.* **1** 80–92

[10] Abdelminaam D S, Ismail F H, Taha M, Taha A, Houssein E H and Nabil A 2021 CoAID-DEEP: an optimized intelligent framework for automated detecting COVID-19 misleading information on Twitter *IEEE Access* **9** 27840–67

[11] Bharti M and Jindal H 2020 Automatic rumour detection model on social media 6th Int. Conf. on Parallel, Distributed and Grid Computing (PDGC) (Piscataway, NJ: IEEE) pp 367–71

[12] Goonathilake M P and KumaraL P V 2020 CNN, RNN-LSTM based hybrid approach to detect state-of-the-art stance-based fake news on social media 20th Int. Conf. on Advances in ICT for Emerging Regions (ICTer) (Piscataway, NJ: IEEE) pp 23–8

[13] Bahad P, Saxena P and Kamal R 2019 Fake news detection using bi-directional LSTM-recurrent neural network *Procedia Comput. Sci.* **165** 74–82

[14] Gupta V, Jain N, Garg H, Jhunthra S, Mohan S, Omar A H and Ahmadian A 2022 Predicting attributes based movie success through ensemble machine learning *Multimedia Tools Appl.* 1–30

[15] Jain N, Jhunthra S, Garg H, Gupta V, Mohan S, Ahmadian A and Ferrara M 2021 Prediction modelling of COVID using machine learning methods from B-cell dataset *Res. Phys.* **21** 103813

[16] Varshney D and Vishwakarma D K 2021 A review on rumour prediction and veracity assessment in online social network *Expert Syst. Appl.* **168** 114208

[17] Varshney D and Vishwakarma D K 2021 A unified approach for detection of clickbait videos on YouTube using cognitive evidences *Appl. Intell.* **51** 4214–35

[18] Varshney D and Vishwakarma D K 2023 An automated multi-web platform voting framework to predict misleading information proliferated during COVID-19 outbreak using ensemble method *Data Knowl. Eng.* **143** 102103

IOP Publishing

Data Analytics for Intelligent Systems
Techniques and solutions
Sachin Taran, Chhavi Dhiman and Manjeet Kumar

Chapter 13

Structural crack detection, segmentation, and classification: a review

**Basavaraj Katageri, Rajashri Khanai, Rajkumar V Raikar,
Dattaprasad A Torse and Krishna Pai**

In the current market, there is phenomenal growth in construction and real estate. The majority of people are likely to build or invest in these civil structures for long-term shelter or investment. However, the major concern that arises is the quality of the building, the construction materials used, and the sustainability and reliability of these civil structures. Early detection of cracks in buildings can increase the life span of the building and avoid any accidents. Hence our chapter presents a detailed review to detect, segment, and classify cracks in buildings. This comparative study attempts to understand the various algorithms, models, and techniques available or being developed by researchers for crack detection. This chapter provides the literature for an experimental review conducted based on the type of algorithm/ technique used, its accuracy and precision, and the type of construction material on which the model is implemented. Over 55 methods and algorithms are explored in this article with respect to the structural material availability.

13.1 Introduction

The construction of huge modern concrete civil structures is currently a popular endeavour. However, maintaining these built structures is highly complex. This is due to the early detrition or damage of buildings due to incorrect maintenance. The building undergoes a lot of wear and tear from the time of construction through the time of occupancy. This wear and tear is mostly caused by incorrect planning, improper curing, incorrect concrete mixtures, and so on, resulting in the formation of cracks on the surface or inside the surface of the built structure [1]. The built structures can be walls, roofs, pavements, underwater structures, or floors.

These cracks may often be divided into two categories: structural cracks and non-structural cracks [2]. Non-structural cracks only have an exterior impact and do not impair the building's stability, whereas structural cracks have the potential to do so. In accordance with the width of the crack, the structural crack is further categorised into three categories: wide, medium, and thin [3], with widths greater than 2 mm, between 1 and 2 mm, and less than 1 mm, respectively. Structural cracks are mainly formed in the columns, slabs, and beams of the structure. Multiple preventive methods are available to stop the further spread of cracks, such as gravity filling, routing, sealing, underpinning, plugging, etc.

Early detection and understanding of the source of the formation of cracks can be highly beneficial to form a systematic solution-based approach to maintaining the structure. The systematic maintenance approach planned thus helps to increase the lifetime of the building to a greater extent. A non-destructive crack-detecting method can be used. The approach can use images or video data of cracks as the input, and the intended result can be an understanding of the type of crack, and its size, intensity, and severity. These fundamental elements aid in identifying the cause of the crack, enabling planning for a remedy to prevent the crack from spreading, and preventing the structure from deteriorating earlier than expected.

This article focuses on understanding and comparing distinct types of machine learning, deep learning, and artificial intelligence-based algorithms. The comparison is done in order to understand the precision, accuracy, and structure materials for the various available and proposed models/algorithms. Figure 13.1 describes the structural building materials considered for surface crack detection under various conditions. A generalised methodology for the rack detection technique is given briefly in section 13.2. The literature review and experimental review are explained in greater detail in section 13.3. Section 13.4 analyses all the potential outcomes and discusses how cracks may occur on materials including asphalt, concrete, steel beams, and mixtures of all of these. In conclusion, our observations and future work are included in section 13.5.

13.2 Methodology

In a general scenario, an image dataset consisting of structural cracked and the uncracked surfaces is used to train the model. Tp replace already existing image datasets, a new dataset can be generated using camera sensors. Figure 13.2 describes a general methodology of how a crack detection model works. Due to advancements in algorithms, datasets other than images can also be used. The sensors in use can be wired or wireless within the system. In the race of technological growth, many implantable sensors inside civil structures are also available [4, 5].

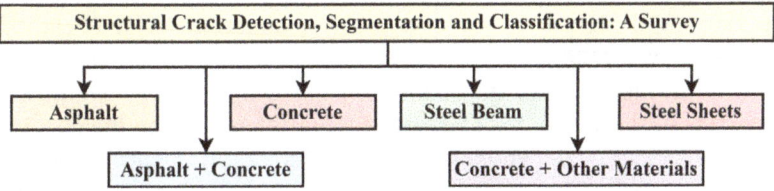

Figure 13.1. Structure materials considered in the review.

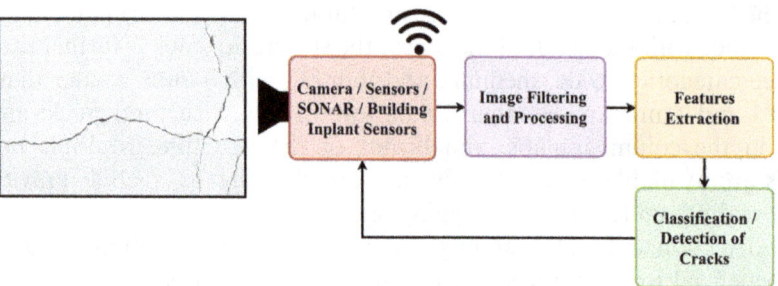

Figure 13.2. Block diagram for crack detection.

The existing or newly generated image datasets contain noise. This noise is due to the lighting conditions, handling of the camera sensors, the material, or the device. This noise must be eliminated before the images are used for training or testing. Failing to do so might result in a dramatic decrease in the accuracy values of the crack detection output. Image filtering and processing can be done by various image processing techniques [6]. This step also helps us to understand the region of interest, the size, and the type of crack. Multiple filters, such as canny edge detector, Shannon, gradient, Gaussian, and median filters, etc, can be used in this process.

Dimensionality reduction is required for larger scale images. Therefore, feature extraction is a crucial step that helps in the reduction of the amount of redundant data from the dataset in use. This process helps in the reduction of the processing time in the next process, thus increasing the efficiency of the model. The final and major step is the detection and classification of internal or external cracks in the civil structure. To do so multiple algorithms in the literature are explored and compared.

13.3 Literature and experimental review

In recent times, while addressing solutions for crack detection, the first solution that crops up in a researcher's mind has been to classify the cracks using crack images of civil structures. This solution can be achieved largely using computer vision technology. The authors in [7] trained and tested two separate datasets consisting of images from surfaces such as asphalt and concrete. Using an architecture called residual convolution neural networks (residual CNN), the authors reached an accuracy of 97.8% for concrete structures and 87.6% for asphalt structures. The authors even explored other learning approaches, such as joint, sequential, and ensemble learning. In the case of joint learning, good accuracies of about 88.4% and 97.8% for asphalt and concrete surfaces were achieved, respectively, while for ensemble learning a high accuracy of 97.7% for concrete surfaces and 82.4% for asphalt surfaces was obtained. Future work should attempt to develop a highly robust system that is independent of the structure material that could detect more than just cracks in the structure. This future work is inspired by the conclusion that crack detection accuracy depends on the type of material. Hence, there could be a significant reduction in the crack detection accuracy if there is a change in the structure's material.

During non-destructive inspection (NDT) or structural health monitoring (SHM), image processing has proved to be an emerging solution. The research in [8] introduces the scale-invariant feature transform (SIFT) method. This matching method is an alternative strategy for the selective elimination of geometric distortion errors and residual reprojections. In this article, a variety of fully convolutional networks (FCNs) were introduced along with feature descriptors to accurately determine the size and location of the cracks on a steel beam. By the end of the experiment, the authors concluded with the five best FCN-based networks. They were successful in achieving high-quality crack detection on a steel beam structure. Future work could be expected to improve the robustness of the FCN-based model, increase the size of the dataset, and lastly combine it with a semi-supervised learning method in the case of a small number of data samples.

With the vast growth in technology, unmanned aerial vehicles (UAVs) are being introduced widely in the inspection of buildings for damage or abnormalities. The UAVs help in capturing images of damage or cracks on the surface of structure where it is difficult for humans to reach. However, the images captured by UAVs contain background noise and quality complications. To tackle this, the authors in [9] introduced a two-step model based on a deep learning algorithm. Façade crack images captured using UAVs were used to train the models. A convolutional neural network (CNN) model was used in the first step. The CNN model was trained using 26 177 images which presented 94% precision. During the second step the model was trained using the U-net neural network model. 96% precision was obtained by training the U-net neural network model with 2870 images. Focusing on façade crack detection using a combination of faster R-CNN and U-Net segmentation models is expected to be the future of this work. By leveraging the power of these two models, it is anticipated that more accurate and efficient façade crack detection can be achieved.

As seen in the above article, to train a deep learning model a user must use an immense amount of images in datasets. The generation or gathering of a large number of images is highly laborious. Moreover, training this large amount of data demands greater computing power. Hence this would be a limitation for the model. To overcome this, the authors in [10] introduced a new framework called KrakN. KrakN is an artificial intelligence-based framework used for dataset generation, training the model, and deploying the model to support infrastructure maintenance. Its architecture is based largely on transfer learning, yielding an accuracy greater than 90%. This framework can also be used on unseen data, thereby supplying a lower metric loss.

Computer vision-based solutions for structural crack detection are restricted to only detecting the cracks on the external surfaces of civil structures. Hence the internal cracks in civil structures cannot be found, thereby limiting the reach of the user. Thermal imaging based crack detection helps us to overcome these limitations by providing us with the internal features of the structure. The authors of [11] devised an innovative method of combining neural networks with thermal imaging. A steel sheet was first excited with heat using a rolling electric heater. Then the data were captured by understanding the thermal (temperature) difference between the cracked and uncracked areas. This process was performed using an infrared thermal imaging process. The captured data were trained and validated using the CNN model. A combination of a faster region based convolutional neural network (faster R-CNN) and region proposal network (RPN) was employed to obtain an accuracy of 95.54%. The future work is to improve the heating performance and understand various morphological crack distributions with respect to various heating methodologies, thereby developing software for real-time detection.

The author of [12] proposed a ResNet101-based end-to-end crack segmentation model. To improve the segmentation accuracy this model integrates many low-level features. The model was trained and compared with FCN using the concrete crack dataset. The highest precisions obtained were 93.12% and 95.69% from FCN and the proposed model, respectively.

Usually, the models designed using deep learning algorithms are expected to be complex and high in cost. Therefore, for low-cost applications, lower complexity and highly flexible are needed. The authors in [13] proposed a model developed based on image processing techniques. The model is largely based on a newly designed strategy used for threshold selection depending on relative standard deviation. Relative standard deviation works on the differences in image segmentation. An anisotropic diffusion process, top hat filtering, and bottom hat filtering processes are used to enhance the image's quality and contrast. The future work is to build a real-time system for videos, pre-processing, and classification of surface cracks on a concrete system.

With limited resources for computing the images for crack detection, the authors in [14] used an existing model to improve its capabilities. The model used was EfficientNetB0. EfficientNetB0 uses a transfer learning method designed using neural architecture search technology. ImageNet was used for the weights. The Adam optimiser was used in supervised learning for updating network parameters. This experiment produced an accuracy of 99.11%. This result was compared with three other existing models. The future work is to develop a system that can characterise levels of crack severity.

In a real-time crack detection scenario, it is difficult to obtain a perfect image of the crack every time. The clarity, lighting conditions, weather conditions, and the camera distance depend directly or indirectly on the quality of the detection accuracy. Therefore the author in [15] experimented with using a pre-trained faster R-CNN model. In this experiment, the lighting conditions, weather conditions, and camera distance were kept as a variable for measurement of accuracy. It was found that all the crack images were detected on a sunny day, whereas on a foggy day most of the crack images were left undetected. Only 15% of cracks could be detected in late evenings and success rate of crack detection decreased to half.

The authors of [16] proposed a full-fledged crack detection and classification model for asphalt pavements. This model was developed based on a deep convolutional neural network (DCNN) architecture. The best part of the model is that it can work on two varied filter sizes. It was observed that the smaller the filter size, the more time was required for training. The crack detection accuracy was around 99% and precision was 99.1%. In the case of crack classification, the model reached an accuracy of 94.5%, considering the classes no crack, alligator, transverse, and longitudinal. Even though the results are promising, the authors faced issues such as the existence of shadows, water spots, irregularities, and oil strains on the pavement. This functioned as noise for detection and classification. Hence further work would be to classify the cracks into a variety of other types and to improve the quality by reducing the pattern noise.

For crack segmentation in the situation of tunnel structures, the authors in [17] proposed a model using mask region based convolutional neural networks (mask R-CNNs) in combination with a morphological closing operation. This model can be used to perform instance segmentation of tunnel cracks using shield tunnel lining images. The mask R-CNN is divided into different architectures. The first is the backbone architecture and the second is the head architecture. The backbone architecture is the region proposal network (RPN), and the head architecture is for specifications. To integrate the disjointed cracks that belong to a single crack, the morphological closing operation is combined with mask R-CNN. The accuracy, for example, of segmentation of tunnel cracks is 81.94%. The enlargement of the database holding multiple types of cracks is required to improve the accuracy and to generalise the model for different environments inside the tunnel, and is expected in future work.

In our study, we present another crack detection model based completely on a image processing technique. Standard Otsu method outcomes based on numerous factors of noise, such as level of illumination, pollution, and contrast in the image are not so satisfactory. Hence, the author in [18] proposed a model based on the green intensity adjustment methodology. This method is also called min–max grey level discrimination (M2GLD). M2GLD is used for the pre-processing of the image threshold by the Otsu method. This proposed model is highly capable of identifying cracks and analysing the parameters of the cracks, such as orientation, area, width, perimeter, and length. The only limitation found in this model is that there are two important parameters that must always be fine-tuned while using the model for better accuracy. These parameters are adjusting the ratio and margin parameters, represented as RA and τ, respectively. As per the author's study, an RA value set to 2 and τ set to 0.5 would deliver a satisfactory outcome. The addition of optimisation methods and self-calculation of the best value for RA and τ would be the focus of future work.

For crack detection in a building consisting of five or more stories, the authors in [19] proposed and implemented a model which captures the crack images using UAVs. The proposed model is built based on the architecture of a modified deep hierarchical CNN. The architecture is 16 convolutional layers combined with a cycle generative adversarial network (CycleGAN). In addition to images captured via UAVs, the authors also use an open-source dataset. The applications of conditional random fields (CRFs) and guided filtering (GF) are employed to obtain remarkably reliable results, thereby refining the outputs produced by the prediction. The major advantage stated by the author of this model is its ability to combine multilevel and multi-scale features during training procedures, and also its applications in reducing noise. The above model is evaluated with five other methods for detailed validation. An accuracy of 99% was obtained using this model.

The authors in [20] proposed a novel approach for multi-directional crack detection. The whole algorithm depends on the Gabor filter producing a precision of around 95%. Gabor filters are a joint domain function chosen to achieve higher analytical resolution and lower bounds in the joint domains. A bank of filters containing multi-orientation filters is generated using the Gabor filter. The future work aims to enhance the existing algorithm by the addition of post-processing techniques and by the analysis of connected components.

The images captured for training and testing deep learning models are usually captured from a high-definition camera sensor. This adds too much to the cost of the model. To overcome this, the authors in [21] introduced and studied a supervised deep convolutional neural network model. This model uses images from low-cost sensor devices, such as smartphones. The outcome of this model was compared with two other models and was found to be comparatively better with a precision of 86.96%. Optimising the existing model and thereby building an integrated system for low-cost real-time crack detection on road surfaces is stated to be the aim of future work.

We highlighted an algorithm better than Otsu's thresholding method in a previous description of an article. Along the same lines, another image segmentation algorithm was proposed by the authors in [22] to produce a precision of 99.92% and an accuracy of 98.7%. This novel model consists of a fully connected neural network and a k-means clustering for image classification and adaptive segmentation, respectively. In the first step, the model detects the presence of cracks on this surface. In the second step, the image is smoothed to minimise the noise pixels with the help of bilateral filtering. In the last step to extract the crack sections from the road surface, an adaptive thresholding methodology is employed. Even though this model produces a higher accuracy and precision percentage, proper segmentation of most of the noisy pixels is not possible.

A detailed comparison experimental study was carried out by the authors in [23]. This experimental study concentrated on the mask R-CNN and faster R-CNN algorithms. The model developed by combining both algorithms using the strategy of joint training is found to be highly effective. This model was trained under five varieties of learning rates. It was also observed that this jointly trained model outperformed YOLOv3. Even with so many advantages, this strategy degrades the effectiveness of the detected bounding box by the mask R-CNN. In the majority of the training and testing, the accuracy was between the range of 96% to 100%. In future work, the authors want to use a small set of label data for more effective results, investigating and improving the existing algorithm to develop a more advanced model for crack detection.

Another report [24] consists of a detailed comparison between three neural network models, namely ResNet18, VGGNet13, and AlexNet. A common training dataset was generated and built for this model. The experiment indicates that ResNet18 produces the best result among the three neural networks. VGGNet13, being more complex due to its substantial number of convolutional layers, outperforms AlexNet. AlexNet has a comparatively simple network. In the second study, it was found that YOLOv3 also produces a remarkably high accuracy level. The author states that their novel work is to combine ResNet18 and YOLOv3 to form a single model, in which the cracks are recognised quickly by ResNet18, and then YOLOv3 detects the crack area from images or videos.

A group of authors [25] planned an institution-wide crack detection inspection. The campus of this institution was made up of a total of seven buildings, whose construction dates vary from 1985 to 2015. The authors' goals included identifying the types and severity of cracks and offering a solution utilising an image processing method. Consequently, an image processing model was created, combining the Otsu image binarization procedure with M2GLD. The cracks' length, width, area, perimeter, and orientation could all be determined using this model.

The authors created an autonomous deep learning-based crack-detecting algo-rithm [26]. On 40 000 photos, edge detection, image thresholding, and other pre-processing techniques were applied. The authors investigated thoroughly whether crack detection depends on RGB or grayscale image types. This study was created using a VGG16 model that was pre-trained. The study revealed that whether images are RGB or grayscale does not impact how well the deep learning model performs when identifying cracks in a concrete building. As a result, the authors claimed that the image's colour is not an important factor in crack identification. Additionally, they claimed that image processing approaches decreased performance.

The authors compared models based on deep learning versus digital image processing (DIP) models in an experimental setting. Four different types of surface crack image datasets were used in this study. CNN models were created for deep learning with and without data pre-processing, whereas for digital image processing a model using the Otsu thresholding method and another model using the adaptive thresholding method were both employed. It was found that the model based on digital image processing designed using the Otsu thresholding approach outper-formed the other four models due to various factors, as listed in [27].

Due to uneven light sources at the sites, many researchers have problems with crack recognition throughout the image processing implementation process. To tackle this issues, the authors suggested a novel image processing strategy as a potential remedy for the problem in [28]. Contrast-limited adaptive histogram equalisation (CLAHE), adaptive image thresholding using first-order local statistics, and noise filtering with nonlinear diffusion filtering were all used. This highly precise and reliable model was primarily suggested for every type of road, tunnel, asphalt pavement, bridge, concrete construction, etc.

The authors in [29] proposed a hybridised strategy to minimise the contamination in images caused by varied types and intensities of noise. This hybridised technique, which produced an accuracy of 91.87%, combines transfer learning, pre-trained CNN, and a modified Dempster–Shafer (D-S) algorithm. The model was trained using 41 780 images of various deformed concrete buildings. To increase the accuracy of crack identification, the D-S algorithm performed the final decision-level image fusion. Additionally, a scanning window based on exhaust search was developed to assess the accuracy of incorrect crack prediction.

The authors' research included a semantic segmentation technique [30]. To increase the model's speed and accuracy, hierarchical CNN and a multi-loss updating approach were used in its development. 1196 images were used to train this model. The primary goal of this model's proposal was to provide a quick and more accurate crack detection system for structural safety inspections. With an increase in the intersection over a union of cracks by 2.165%, the average inference time dropped to roughly 65.90%.

Despite requiring a lot of resources, traditional image processing methods fall short of the desired degree of accuracy. The authors recommended a modified U-Net architecture as a solution to this problem in [31]. In this architectural design, dice loss was used as the evaluation's objective function, residual blocks were used instead of all the convolutional layers, and AMSGrad optimisation was used to measure performance. The test was run on a variety of datasets, and the authors were able to obtain state-of-the-art performance, but only at the expense of labour-intensive manual annotations of the data. The CLAHE or histogram equalisation (HE) approach was recommended by the authors for uneven lighting conditions.

13.4 Results and discussion

The above literature and experimental review can be divided into three main parts: asphalt and concrete based structures, concrete with other materials, and steel-based structures. Table 13.1 describes various algorithms that can be used on asphalt and concrete based structures with their respective accuracy and precision.

When considering structures developed purely out of asphalt materials, crack detection can be carried out easily using a neural network architecture such as CNN [7], DCNN [16], joint learning [7], and ensembled learning [7]. In this review it was observed that when the DCNN algorithm is used in crack detection the accuracy and precision are higher compared to other algorithms. This was due to the algorithm's ability to accurately identify the important features from the images automatically. However, it requires a large volume of image datasets and is unable to encode the orientation of the desired objects from the image [32].

Many new algorithms and techniques were considered when the same analysis was conducted using only pure concrete buildings. Algorithms such as VGG16 [26], EfficientNetB0 [14], and DenseNet201 [14] outperformed when compared with other algorithms. The metrics considered for this discussion were mainly the percentage of accuracy and precision as mentioned in the respective author's work. In comparison to the other two top models, EfficientNetB0 has the smallest size, fastest inference time (on CPU), and maximum accuracy with the fewest parameters, making it the superior model [33]. Due to its capacity to improve its performance metrics with the use of multiple filters and techniques, VGG16 can be placed second. It was also observed that M-R-50-C4 [17] performed the worst when compared to other algorithms.

Similar studies were also considered to understand the best possible algorithm that can be used for structures developed using a mixture of concrete and asphalt. The outcome shows that DeepCNN with adaptive thresholding [22], VGGNet13 [24], and

Table 13.1. Various algorithms used on asphalt and concrete based structures.

Sr. no.	Algorithm/techniques used	Accuracy	Precision	Materials used	References
1	Residual CNN	87.60%	—	Asphalt	[7]
2	Joint learning using a 12-layer model	88.40%	—	Asphalt	[7]
3	Ensemble learning using a 12-layer model	82.40%	—	Asphalt	[7]
4	DCNN (detection)	99%	99.10%	Asphalt	[16]
5	DCNN (classification)	94.50%	—	Asphalt	[16]
6	Residual CNN	97.80%	—	Concrete	[7]
7	Joint learning using a 12-layer model	97.80%	—	Concrete	[7]
8	Ensemble learning using a 12-layer model	97.70%	—	Concrete	[7]
9	ResNet101-based end-to-end crack segmentation model	—	95.69%	Concrete	[12]
10	FCN	—	93.12%	Concrete	[12]
11	Relative standard deviation	—	93%	Concrete	[13]
12	MobileNetV2	97.82%	98.21%	Concrete	[14]
13	EfficientNetB0	99.11%	98.78%	Concrete	[14]
14	DenseNet201	99.32%	98.92%	Concrete	[14]
15	InceptionV3	98.98%	98.91%	Concrete	[14]
16	M-R-50-C4	68.19%	—	Concrete	[17]
17	M-R-50-FPN	79.35%	—	Concrete	[17]
18	M-R-101-FPN	79.10%	—	Concrete	[17]
19	M-R-101-FPN-closing-operation	81.94%	—	Concrete	[17]
20	16conv + CycleGAN + guided filter	99%	83.80%	Concrete	[19]
21	Baseline method	98.80%	84%	Concrete	[19]
22	DeepCrack-BN	98.20%	76.80%	Concrete	[19]
23	DeepCrack-GF	96.40%	78.70%	Concrete	[19]
24	SegNet	87.10%	62.60%	Concrete	[19]
25	Mask R-CNN + faster R-CNN	96%> to < 100%	—	Concrete	[23]
26	VGG16 + RGB images	99.53%	—	Concrete	[26]
27	VGG16 + grayscale images	99.550%	—	Concrete	[26]
28	VGG16 + OTSU method	98.817%	—	Concrete	[26]
29	VGG16 + Sobel filter	99.13%	—	Concrete	[26]
30	CNN + data pre-processing	49%	—	Concrete	[27]
31	CNN	33.5%	—	Concrete	[27]
32	DIP + Otsu thresholding technique	93.22%	—	Concrete	[27]
33	DIP + adaptive thresholding technique	85.59%	—	Concrete	[27]
34	Pre-trained CNN + transfer learning + modified Dempster–Shafer (D-S) algorithm	91.87%	85.58%	Concrete	[29]
35	U-Net architecture + dice loss + AMSGrad + residual blocks with Crack500 dataset	—	80.95%	Concrete	[31]

36	U-Net architecture + dice loss + AMSGrad + residual blocks with DeepCrack dataset	—	85.35%	Concrete	[31]
37	Gabor filter	—	95%	Concrete and asphalt	[20]
38	Support vector machine (SVM)	—	81.12%	Concrete and asphalt	[21]
39	Boosting method	—	73.60%	Concrete and asphalt	[21]
40	ConvNets	—	86.96%	Concrete and asphalt	[21]
41	DeepCNN + adaptive thresholding	98.70%	99.92%	Concrete and asphalt	[22]
42	Otsu's thresholding	98.48%	95.90%	Concrete and asphalt	[22]
43	AlexNet	97.60%	97.90%	Concrete and asphalt	[24]
44	VGGNet13	98.30%	98.60%	Concrete and asphalt	[24]
45	ResNet18	98.80%	98.90%	Concrete and asphalt	[24]

ResNet18 [24] provided a good percentage of accuracy and precision over other algorithms. Even though VGGNet13 offers superior accuracy compared to ResNet18, the addition of adaptive thresholding further enhances any deep CNN model by improving segmentation and reducing the misclassification probability [34]. Table 13.2 describes various algorithms that can be used on the structures developed using concrete and building materials with their respective accuracy and precision.

From table 13.2, compared to other algorithms, KrakN [10] and CrackNet [10] produced results with a higher accuracy percentage. The proposed two-step model from the author in [9] gave a greater degree of precision through the application of pixel-level analysis and reduction of noise regions in the input images. The synergistic combination of patch- and pixel-level detection methods outperformed other techniques in its category. This innovative approach has proven to be a powerful

Table 13.2. Various algorithms used on the structures developed using concrete and other building materials.

Sr. no.	Algorithm/ Techniques used	Accuracy	Precision	Materials used	References
1	KrakN + KrakN dataset	98%	—	Concrete, brick, stone, metal, wood, etc	[10]
2	CrackNet + KrakN dataset	48%	—	Concrete, brick, stone, metal, wood, etc	[10]
3	DeepCrack + KrakN dataset	26%	—	Concrete, brick, stone, metal, wood, etc	[10]
4	KrakN + CrackNet dataset	81%	—	Concrete, brick, stone, metal, wood, etc	[10]
5	CrackNet + CrackNet dataset	98%	—	Concrete, brick, stone, metal, wood, etc	[10]
6	DeepCrack + CrackNet dataset	97%	—	Concrete, brick, stone, metal, wood, etc	[10]
7	Convolutional neural network (CNN)	—	94%	Concrete/cement, brick, stone, polymer, glass, metal, rammed earth, wood	[9]
8	U-Net neural network model	—	96%	Concrete/cement, brick, stone, polymer, glass, metal, rammed earth, wood	[9]
9	Proposed 2-step CNN + U-Net model	—	97.40%	Concrete/cement, brick, stone, polymer, glass, metal, rammed earth, wood	[9]

tool for achieving superior results. Table 13.3 describes various algorithms that can be used on the structures developed using steel materials with their respective accuracy and precision.

The author in [8] used a variety of algorithms for steel beam constructions in order to locate and section cracks in them. On average, these algorithms' precision results are comparable. Table 13.3 summarises the same. Of these, the structural forests with wavelet transform (SFW)-hrbio1.1 algorithm outperformed its counterparts, despite its intensive computation requirements. Its remarkable performance was a testament to its efficiency and efficacy.

A similar study was carried out by another author [11], in which steel sheets were considered. The improved faster R-CNN (+ RPN) technique performed with a higher proportion of percentage accuracy and precision. This was due to its capability to simultaneously predict object scores and boundaries at a specific

Table 13.3. Various algorithms used on structures developed using steel materials.

Sr. no.	Algorithm/Techniques used	Accuracy	Precision	Materials used	References
1	FCN-8s-Conv4-Conv5	—	63.51% and 61.37%	Steel beam	[8]
2	FCN-8s-Conv5	—	59.94% and 61.83%	Steel beam	[8]
3	FCN-8s	—	57.84% and 60.34%	Steel beam	[8]
4	FCN-8s+2conv	—	56.31% and 55.36%	Steel beam	[8]
5	FCN-8s+2conv+ Conv6	—	54.70% and 53.51%	Steel beam	[8]
6	FCN-8s+2conv+ Conv6+Dec4	—	58.76% and 58.75%	Steel beam	[8]
7	Structural forests with wavelet transform (SFW)-hrbio1.1	—	62.62% and 61.46%	Steel beam	[8]
8	Fast edge detection using structured forests (SFD)	—	61.70% and 59.84%	Steel beam	[8]
9	DeeplabV3	—	54.34% and 53.36%	Steel beam	[8]
10	ResNet-152	—	54.17% and 52.95%	Steel beam	[8]
11	ResNet-50	—	51.70% and 52.60%	Steel beam	[8]
12	HOG-SVM	86.62%	74.62%	Steel sheets	[11]
13	YOLOv3	92.15%	87.12%	Steel sheets	[11]
14	Original faster R-CNN	92.36%	90.53%	Steel sheets	[11]
15	Improved faster R-CNN (+ RPN)	95.54%	92.41%	Steel sheets	[11]

position, making it a highly effective technique [35]. Table 13.4 summarises the best algorithms reviewed with respect to the construction materials used.

Non-destructive evaluation methods including computer vision, image processing, and machine learning are in high demand due to the development of technology. Depending on the size and orientation of the cracks, various techniques can be effectively used to identify them with very few resources. The probability of a correct prediction or identification, however, is frequently a key factor in these systems. Consequently, they are less cost-effective and lack scalability for larger areas [36].

Table 13.4. Possible best algorithms with their construction materials.

Sr. no.	Materials used	Best algorithms/techniques
1	Asphalt	DCNN [16]
2	Concrete	EfficientNetB0 [14] or VGG16 [26]
3	Asphalt + concrete	DeepCNN + adaptive thresholding [22]
4	Concrete + other materials	Proposed 2-step CNN + U-net model [9]
5	Steel beam	Structural forests with wavelet transform (SFW)-hrbio1.1 [8]
6	Steel sheets	Improved faster R-CNN (+ RPN) [11]

13.5 Conclusion

This chapter provides us with more than 55 varieties of models, algorithms, and techniques for crack detection and classification. This detailed study can assist researchers to understand which model or technique should be used concerning the structural material availability. This study also helps us to improve existing models by understanding the limitations and intended future work of researchers. In the current trending scenarios, models such as machine learning, deep learning, and artificial intelligence can be combined with image processing techniques to obtain a highly precise and accurate output. These improvements can also reduce the training times of the models drastically, thereby also reducing the time for computation. The study concludes by stating a few of the best possible algorithms that can be used for implementation with respect to structural material availability. The aim of future work would be to implement these selected best models in the real-time scenarios and validate the same. In consequence, the life expectancy of buildings could be increased drastically through proper pre-planned maintenance. This, in turn, avoids accidents caused by damage to civil structures.

References

[1] K C D and Sudhakar D V K 2020 Deterioration of a building through environmental and anthropological causes *Int. J. Eng. Adv. Technol.* **9** 2681–96
[2] De Thales A K J and Sadeghi K 2022 Causes and effects of structural cracks *Int. J. Mod. Trends Sci. Technol.* **8** 64–9
[3] Semwal S, Gangwar P and Bahuguna A 2020 Review paper on cracks in building and their remedies *Int. Res. J. Eng. Technol.* **7** 5010–2
[4] Chakraborty J, Katunin A, Klikowicz P and Salamak M 2019 Early crack detection of reinforced concrete structure using embedded sensors *Sensors* **19** 1–22
[5] Stałowska P, Suchocki C and Rutkowska M 2022 Crack detection in building walls based on geometric and radiometric point cloud information *Autom. Constr.* **134** 104065
[6] Cao J 2021 Research on crack detection of bridge deck based on computer vision *IOP Conf. Ser.: Earth Environ. Sci.* **768** 012161

[7] Alipour M and Harris D K 2020 Increasing the robustness of material-specific deep learning models for crack detection across different materials *Eng. Struct.* **206** 110157

[8] Wang S, Liu C and Zhang Y 2022 Fully convolution network architecture for steel-beam crack detection in fast-stitching images *Mech. Syst. Signal Process.* **165** 108377

[9] Chen K, Reichard G, Xu X and Akanmu A 2021 Automated crack segmentation in close-range building façade inspection images using deep learning techniques *J. Build. Eng.* **43** 102913

[10] Żarski M, Wójcik B and Miszczak J A 2021 KrakN: transfer learning framework and dataset for infrastructure thin crack detection *SoftwareX* **16** 100893

[11] Yang J, Wang W, Lin G, Li Q, Sun Y and Sun Y 2019 Infrared thermal imaging-based crack detection using deep learning *IEEE Access* **7** 182060–77

[12] Meng X 2021 Concrete crack detection algorithm based on deep residual neural networks *Sci. Program* **2021** 3137083

[13] Peng T, Kavya T S, Jang Y M and Kim B W 2020 Concrete crack detection using relative standard deviation for image thresholding *Allergy, Asthma Immunol. Res.* **13** 2720–8

[14] Su C and Wang W 2020 Concrete cracks detection using convolutional neuralnetwork based on transfer learning *Math. Probl. Eng.* **2020** 7240129

[15] Hacıefendioğlu K and Başağa H B 2022 Concrete road crack detection using deep learning-based faster R-CNN method *Iran. J. Sci. Technol. Trans. Civ. Eng.* **46** 1621–33

[16] Yusof N A M, Ibrahim A, Noor M H M, Tahir N M, Yusof N M, Abidin N Z and Osman M K 2019 Deep convolution neural network for crack detection on asphalt pavement *J. Phys. Conf. Ser.* **1349** 012020

[17] Huang H, Zhao S, Zhang D and Chen J 2022 Deep learning-based instance segmentation of cracks from shield tunnel lining images *Struct. Infrastruct. Eng.* **18** 183–96

[18] Hoang N-D 2018 Detection of surface crack in building structures using image processing technique with an improved otsu method for image thresholding *Adv. Civ. Eng.* **2018** 1–10

[19] Munawar H S, Ullah F, Heravi A, Thaheem M J and Maqsoom A 2021 Inspecting buildings using drones and computer vision: a machine learning approach to detect cracks and damages *Drones* **6** 5

[20] Salman M, Mathavan S, Kamal K and Rahman M 2013 Pavement crack detection using the Gabor filter 16th Int. IEEE Conf. on Intelligent Transportation Systems (ITSC 2013) (Piscataway, NJ: IEEE) pp 2039–44

[21] Zhang L, Yang F, Daniel Zhang Y and Zhu Y J 2016 Road crack detection using deep convolutional neural network 2016 IEEE Int. Conf. on Image Processing (ICIP) (Piscataway, NJ: IEEE) pp 3708–12

[22] Fan R, Bocus M J, Zhu Y, Jiao J, Wang L, Ma F, Cheng S and Liu M 2019 Road crack detection using deep convolutional neural network and adaptive thresholding *IEEE Intelligent Vehicles Symp. (IV)* vol 2019 (Piscataway, NJ: IEEE) pp 74–9

[23] Xu X, Zhao M, Shi P, Ren R, He X, Wei X and Yang H 2022 Crack detection and comparison study based on faster R-CNN and mask R-CNN *Sensors* **22** 1215

[24] Yang C, Chen J, Li Z and Huang Y 2021 Structural crack detection and recognition based on deep learning *Appl. Sci.* **11** 2868

[25] Velumani P, Mukilan K, Varun G, Divakar S, Muhil Doss R and Ganeshkumar P 2020 Analysis of cracks in structures and buildings *J. Phys.: Conf. Ser.* **1706** 012116

[26] Golding V P, Gharineiat Z, Munawar H S and Ullah F 2022 Crack detection in concrete structures using deep learning *Sustainability* **14** 8117

[27] Yadhunath R, Srikanth S, Sudheer A, Jyotsna C and Amudha J 2022 Detecting surface cracks on buildings using computer vision: an experimental comparison of digital image processing and deep learning Soft Computing and Signal Processing (Singapore: Springer) pp 197–210

[28] Parrany A M and Mirzaei M 2022 A new image processing strategy for surface crack identification in building structures under non-uniform illumination *IET Image Process* **16** 407–15

[29] Yu Y, Samali B, Rashidi M, Mohammadi M, Nguyen T N and Zhang G 2022 Vision-based concrete crack detection using a hybrid framework considering noise effect *J. Build. Eng.* **61** 105246

[30] Kim J, Shim S, Cha Y and Cho G-C 2021 Lightweight pixel-wise segmentation for efficient concrete crack detection using hierarchical convolutional neural network *Smart Mater. Struct.* **30** 045023

[31] Ghosh S, Singh S, Maity A and Maity H K 2021 CrackWeb: a modified U-Net based segmentation architecture for crack detection *IOP Conf. Ser.: Mater. Sci. Eng.* **1080** 012002

[32] Alzubaidi L, Zhang J, Humaidi A J, Al-Dujaili A, Duan Y, Al-Shamma O, Santamaría J, Fadhel M A, Al-Amidie M and Farhan L 2021 Review of deep learning: concepts, CNN architectures, challenges, applications, future directions *J. Big Data* **8** 53

[33] Keras 2022 Keras applications *Keras Release 2.11.0*

[34] Liu K, Yang J, Sun C and Chi H 2021 Supervised adptive threshold network for instance segmentation arXiv: 2106.03450

[35] Ren S, He K, Girshick R and Sun J 2015 Faster R-CNN: towards real-time object detection with region proposal networks arXiv: 1506.01497

[36] Downey A, D'Alessandro A, Ubertini F and Laflamme S 2018 Automated crack detection in conductive smart-concrete structures using a resistor mesh model *Meas. Sci. Technol.* **29** 035107

IOP Publishing

Data Analytics for Intelligent Systems
Techniques and solutions
Sachin Taran, Chhavi Dhiman and Manjeet Kumar

Chapter 14

A systematic review of fault detection in hardware and software systems

Mayank Yadav and Ruchika Malhotra

Fault detection is crucial in industry. Early discovery of faults may aid in the prevention of subsequent abnormal events. Fault detection can be achieved in a variety of ways. This chapter will cover the fundamental approaches. Currently, methods for finding flaws faster than the customary time restrictions are required. Detection methods include data and signal approaches, process model based methods, and knowledge based methods. Some treatments need very precise models. Early issue discovery increases life expectancy, enhances safety, and lowers maintenance costs. When choosing a flaw detection system, several factors must be considered. Principal component analysis can help find flaws in large-scale systems. Signal models are used when difficulties arise as a result of process changes. This chapter includes a systematic review of the literature and highlights a selection of noteworthy applications. In this chapter we cover two different real-world scenarios that employ two different defect detection methodologies. In other words, we will look at both hardware and software concerns. The first case considers electrical fault detection and a decision tree technique is utilized to detect these defective lines. The algorithm categorizes electrical cables as defective or non-faulty whenever possible. In the second scenario, we shall employ the ensemble learning learning technique to discover faults in each dataset. We will work on the 'camel' dataset (Eclipse repository).

14.1 Introduction

In our changing world, there is increasing concern regarding safety procedures [1]. The most typical method is to improve the existing detection and prediction mechanism. We aim to concentrate on the wear and tear elements when discussing the hardware situation, which includes physical and hardware-related issues [2].

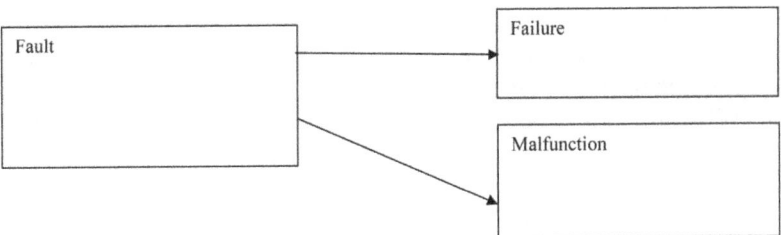

Figure 14.1. A block diagram showing the relationship that exists between a fault, a failure, and a malfunction.

However, if you are considering software flaws, you might not even consider the notion of wearing. Regarding where flaws originate, we can investigate the various methods used for fault detection and their mechanisms [3]. Several faults are observed in our day-to-day lives. To describe the term 'fault', one may say that an undesirable condition of any system is likely to be because of a fault [4].

A defect occurs when a system deviates from the expected behavior or outcome [5]. This chapter will cover two real-world scenarios that employ two different defect detection methodologies. In other words, we will look at both hardware and software concerns. In the first case, electrical fault detection is considered and a decision tree technique is utilized to detect these faulty lines wherever it is suitable to categorize the electrical lines as faulty or non-faulty.

Figure 14.1 depicts the link between a defect and a failure. It also depicts the connection between a problem and a malfunction. To understand things even better, it can be stated that faults are the causes of any failure [6], and when these failures become a recurring phenomenon we recognize them as malfunctions.

14.2 Literature review

14.2.1 Data based and signal model based fault detection

Fault detection can be achieved with the help of some data methods and signal models. These are utilized when there is a chance that we will have access to past data. This historical data might also include experimental data. When only the observed experimental data are utilized to learn about fault detection, we are dealing with data based fault detection [7]. These strategies have advantages, a they are, perhaps, simpler to deal with. This might be due to their overall simplicity. However, an issue occurs when the distribution of data changes. In other words, normally distributed data suits these fault detection methods. If the distribution is something other than normal, then data based methods face some difficulty and might not be the smarter choice [8].

Principal component analysis (PCA) is a well-known data based approach for defect discovery. PCA is employed as a data analysis tool in fault detection. It is commonly utilized in large-scale industry systems. The purpose is to collect a set of observations free of correlation. Transformations are used to convert correlated variables (observations) into uncorrelated variables. This procedure tends to reduce the size of the observations. The uncorrelated data collected thus are referred to as principal components [9]. These have fewer dimensions than the last connected data collection. When the dimensions are decreased, the size of the data decreases. The fewer data there are to handle, the easier they are to manage. Fault detection becomes easier when the observation set is small [10].

Signal models are mathematical models that perform signal analysis when the detection of faults can be linked to the signal-change concept. These models are fundamental mathematical models. The dataset in this category is more likely to have a wave-like structure, which means that columns in this dataset will have characteristics such as amplitude, phase difference, wavelength, and so on. Signal analysis requires the use of the Fourier transform concept, such as the discrete-time Fourier transform, fast Fourier transform, continuous Fourier transform, etc. A commonly known method in this category is the parametric signal model. This model is mainly concerned with frequency changes [11]. These frequency fluctuations are used in the defect detection procedure. When compared to signal-based approaches, there are certain methods that are more adaptable.

Pattern recognition is one such approach. This is based on the idea of the neural network. Their importance grows when the structure of the pattern of the input–output data is complicated or ambiguous. Through the help of neural networks, we represent the situations a system can be in, such as faulty, non-faulty, and under maintenance [12]. These situations are considered conditions which are then represented in the neural networks with the help of 'neurons'.

14.2.2 Process model based fault detection

Fault detection can be achieved with the help of some process models. These are utilized when it is possible to achieve matching outcomes from two separate scenarios. First, we attempt to obtain the real outputs from the system under discussion. The outputs thus obtained are termed system outputs. Second, we take advantage of certain mathematical models [13]. The outputs we can obtain from them are called model outputs [14]. We now observe both outputs and attempt to perform a mapped observation. The resulting mapping is then used to detect problems in a system. The deviation from typical system behavior reveals the presence of flaws inside the system.

Parameter estimation techniques are used to work on the detected input signal data that have been actuated and sensed from the external world. Certain parameters change when a system is infiltrated with defects and becomes defective. We may also use parametric equations to compare present system behavior (as observed by parameter changes) to typical system behavior. These parametric equations are also called parity equations. The main idea of such a fault detection method is to observe the change in the state variables [15]. However, output problems will occasionally indicate malfunctioning software.

14.2.3 Knowledge model based fault detection

Fault detection can be achieved with the help of some knowledge models. These are used when there is a possibility that we may obtain some undivided assistance from different artificial intelligence strategies [16]. First, we attempt to obtain genuine knowledge of the system under discussion. The physical qualities are examined. Second, we use production rules to assist us, from which we obtain 'knowledge'. We now observe an inference engine, a knowledge acquisition based expert system. Rule based expert systems serve as easy domains for fault detection activity [17].

The knowledge acquisition process uses these production rules. We can easily manipulate these rules, i.e. rules can be added and removed as per our will. The concept of fuzzy logic is well-known and is one of the knowledge based methods. Any common fault detection system would simply have to detect two states—faulty and non-faulty [18].

Certain types of signaling mechanisms could indicate faults within a system. One such signaling mechanism is the alarm system. This is like an alarm which is off only when faults are absent, or beeps when a fault is detected. A fuzzy controller is a complex alarm system. It utilizes rule based inferences and processes such as fuzzification and defuzzification.

In the fuzzification process, membership functions graphically represent the input signal magnitudes. These membership functions fuzzify the input data being made available to the fuzzy controllers. In defuzzification, the reverse process is performed [19]. The aim is to obtain the output data back from the fuzzified data.

14.3 Methodology and implementation

14.3.1 Hardware fault detection (faulty electrical lines)

The process of detecting faults and training models to categorize data using a wide range of classifiers can be used to evaluate lines that may have a problem later [20]. Here, we will use the common decision tree algorithm to detect faults in electrical transmission lines. We may conceive of this approach as a supervised machine learning algorithm that executes the learning task by using matched input–output pairs. In our case, the decision tree algorithm will work on an electrical transmission line dataset (VSB power line). It comprises both symmetric and asymmetric faults. The goal is to initially view the data before performing data investigation. The following stage will be to create a machine learning model. This model will then help us in our fault detection objective by predicting transmission lines that may or may not have any faults [21].

A sample snapshot of the original dataset will be shown. The question could arise whether we need to pre-process the data or not, however, pre-processing is necessary and need to be completed [22]. This is done because we are trying to see a modified and simplified version of the dataset [23]. The decision tree classifier was employed in this case. However, a variety of classifiers and regressors might

```
In [1055]:  data.head()
```

Out[1055]:

	G	C	B	A	Ia	Ib	Ic	Va	Vb	Vc
0	1	0	0	1	-151.291812	-9.677452	85.800162	0.400750	-0.132935	-0.267815
1	1	0	0	1	-336.186183	-76.283262	18.328897	0.312732	-0.123633	-0.189099
2	1	0	0	1	-502.891583	-174.648023	-80.924663	0.265728	-0.114301	-0.151428
3	1	0	0	1	-593.941905	-217.703359	-124.891924	0.235511	-0.104940	-0.130570
4	1	0	0	1	-843.663617	-224.159427	-132.282815	0.209537	-0.095554	-0.113983

Figure 14.2. VSB power line dataset.

have been utilized. The accuracy of a machine learning classifier or regressor heavily depends on the quality of the dataset it's trained on. A faulty or non-faulty dataset can impact its performance in differentiating the groups. A machine learning classifier might function in these two groups. When we want the output to be separate categories, classification with machine learning classifiers is required [24]. When it comes to the classifier utilized here, it is the best option for making any type of decision.

Figure 14.2 presents the VSB power line dataset.

The columns in this table of figure 14.2 are as follows:

1. G—stands for 'ground'. It tells us about the line which transfers the current into the Earth's surface.
2. C—stands for 'line C'. It tells us that there is a transmission line with the name C.
3. B—stands for 'line B'. It tells us that there is a transmission line with the name B.
4. A—stands for 'line A'. It tells us that there is a transmission line with the name A.
5. Ia—stands for 'current in line A'. It tells us how much current can be made to flow when we are taking line A into consideration.
6. Ib—stands for 'current in line B'. It tells us how much current can be made to flow when we are taking line B into consideration.
7. Ic—stands for 'current in line C'. It tells us how much current can be made to flow when we are taking line C into consideration.
8. Va—stands for 'voltage in line A'. It tells us how much voltage can be present when we are taking line A into consideration.
9. Vb—stands for 'voltage in line B'. It tells us how much voltage can be present when we are taking line B into consideration.
10. Vc—stands for 'voltage in line C'. It tells us how much voltage can be present when we are taking line C into consideration.

Figure 14.3 shows the dataset visualization of the VSB power line dataset in terms of the feature columns present. The vertical axis represents the number of instances.

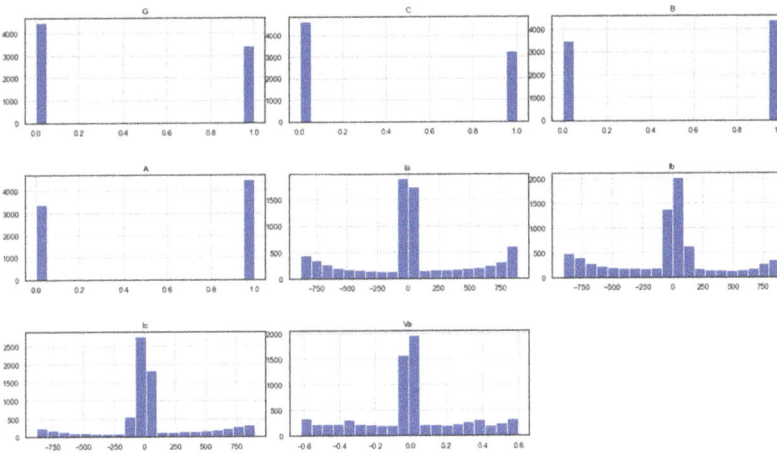

Figure 14.3. Dataset visualization of the VSB power line dataset.

The horizontal axis represents the feature value. The variations shown in different graphs are histograms depicting the feature value graphically.

The following are the steps to take in order to implement the algorithm:

1. Import data files available in the directory. These files contain the dataset which will be used to classify the faults present in the transmission lines.
2. Explore the dataset carefully with the help of the various data visualization techniques available.
3. Build from a variety of machine learning models (here, the decision tree); this will help in the prediction (or detection) process.
4. Import the required libraries.
5. Plot the voltage graph (or current graph). If there is any fluctuating point in any transmission line then it will be detected. Box plots and histograms are commonly suitable in these types of cases.
6. We want to find separate categories, i.e. faulty and non-faulty. Feature separation is the required step. One may also require separation based on labels.
7. Perform the training and testing split.
8. Select the models (here, the decision tree) to generate results. Things such as accuracy score and confusion matrix could also be required.
9. Plott the decision tree. We can now begin comparison if we do this with other models as well.
10. The model can now be made to perform prediction.

14.3.2 Software fault detection

The software fault detection process helps the software testing individuals to enhance the process of finding faults through a diverse set of publicly available fault data, and to examine the dataset effectively and figure out which method could

In [4]: data.head()

Out[4]:

	wmc	dit	noc	cbo	rfc	lcom	ca	ce	npm	lcom3	...	dam	moa	mfa	cam	ic	cbm	amc	max_cc	avg_cc	bug
0	5	3	0	7	10	0	1	7	4	0.250000	...	1.0	1	0.921053	0.360000	1	2	7.400	1	0.6000	0
1	4	1	0	3	5	4	1	2	3	0.666667	...	1.0	1	0.000000	0.500000	0	0	3.000	1	0.5000	0
2	20	4	0	26	95	144	2	26	13	0.842105	...	1.0	0	0.727273	0.197368	4	5	20.300	3	1.0000	0
3	3	2	0	8	22	3	2	6	2	2.000000	...	0.0	0	0.750000	0.666667	1	3	54.000	15	5.3333	1
4	8	1	0	25	20	22	22	3	6	0.571429	...	1.0	0	0.000000	0.250000	0	0	20.875	1	0.7500	1

5 rows × 21 columns

Figure 14.4. The camel dataset.

be utilized to detect (and predict) faults at an earlier stage [25]. In this case, we will use the ensemble learning approach to find flaws in each dataset. To begin with, we may conceive it as a machine learning technique that executes the learning task by boosting individual learner models [26]. The aim is to combine the effects of some learner models to overcome their downsides. We will work on the 'camel' dataset (Eclipse repository) in our second case. It is made up of numerous feature columns. The goal is to prioritize the characteristics before training on them [27]. The following stage will be to create a machine learning model. This model will then aid us in our fault detection goal. There are various performance measures available that can help us determine how efficient our learning ensemble has become. Bagging is a type of bootstrapping aggregation. It is an ensemble learning technique commonly utilized for regressor models [28]. However, bagging is beneficial for classification models as well. In our situation, we used a bagging regressor. Its goal was to reduce the bias–variance tradeoff created by individual regressors such as decision tree (DT), naive Bayes (NB), and logistic regression (LR). Following that, random selection created various training sets to work with. The feature that provides the finest split criteria is constantly sought. Following the acquisition of this feature, a split is conducted to obtain training and test data [29]. The model's accuracy will likely improve when the tree develops through this technique. Figure 14.4 shows the camel dataset and its various characteristic feature columns with their names and values in tabular form.

We can see that there are numerous feature columns present here. They are briefly discussed as follows:

1. Weighted methods per class (WMC) tells us how many methods are there in every class.
2. Depth of inheritance tree (DIT) tells us how many levels of inheritance there are inside a structure of class.
3. Number of children (NOC) tells us how many child classes there are for any parent class.
4. Coupling between object classes (CBO) tells us how many classes a particular class is coupled with.

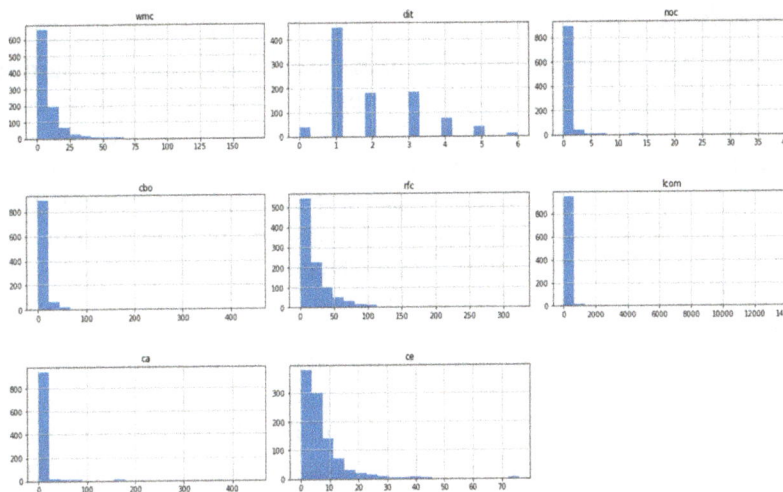

Figure 14.5. Dataset visualization of various metrics present in the dataset description.

5. Response for a class (RFC) tells us how many methods can be made to execute when a message is to be responded to.
6. Lack of cohesion of methods (LCOM) tells us how much methods are correlated to each other.

Figure 14.5 presents dataset visualization to obtain a visual representation of the various metrics present in the dataset description. The vertical axis represents the number of instances. The horizontal axis represents the feature value. The variations shown in different graphs are histograms depicting the feature value graphically.

To obtain the required results, the following steps were taken:

1. Importing the libraries in a 'Jupyter Notebook'.
2. Reading the dataset.
3. Displaying the dataset.
4. Visualizing the dataset .
5. Generating the correlation matrix.
6. Performing data pre-processing to look for null values.
7. Dividing the dataset into training and testing data.
8. Performing the train–test split.
9. The ensemble learning models can now be worked upon.
10. Performing cross-validation (ten-fold).
11. Finally, calculating the accuracy score.

14.4 Results and outcomes

In this section, we present the results of both parts of our study: the hardware and software fault detection methods. We compared the electrical transmission lines based on the presence of faults among them, be it ground faults or line faults. We also calculated the exact count and its respective percentage to denote the presence of faults in the transmission lines; as per our calculations, ground faults were present, and they comprised at least 40% of faulty lines. In contrast, more than 50% of transmission lines were non-faulty. When calculating the faults, we found that as the voltage increases across the transmission lines, not only does the corresponding line load jump, but the fault proneness increases significantly.

The suggested methods aim to develop efficient fault detection. First, the electrical transmission lines A, B, C, and ground are given suitable names. The faults are then checked within these transmission lines to determine which lines are faulty and which are not. Several faults have been evaluated and are shown with the help of figures.

The ensemble learners are used in the second fault detection approach to enhance fault detection rates. A basic learner must be chosen for this. In our experiments this was chosen for three reasons:

1. They outperformed previously studied learners.
2. They are much faster to build than previously used trainers.
3. It is preferable that the base learners are learning algorithms highly affected by training data changes.

The number of iterations in ensemble learners may affect their performance significantly. As a result, each ensemble is tested using a different number of iterations ranging from 10 to 100 iterations, with a step of 10 iterations increasing in each experiment. For all datasets, 10 iterations yielded the best performance [30].

Figure 14.6 shows the data visualization of faulty transmission lines depicting ground faults, and faults in lines A and B. The percentage of faults is shown with the help of a pie chart. A graph is also drawn to depict the respective number of faults present in each transmission line route. '1' denotes fault presence. '0' denotes non-faulty transmission. 'Count' denotes the exact number of faults.

Figure 14.7 shows the voltage distribution visualization along with line load distribution. Figures 14.8(a)–(c) show the load distribution prevalent in power lines A, B, and C, respectively.

The decision tree algorithm produced the results are shown in figure 14.6. It has been shown here, through the help of data visualization, what percentage of faults are present on a particular stretch of electrical transmission lines [31]. In other words, 30% of the lines are operational, and no defects have been identified thus far.

The electrical transmission lines were checked for faults by considering one line at a particular time. Also, visualizing the voltage distribution along the line with its load distribution is mentioned important in existing research. This has been done with the help of graphs.

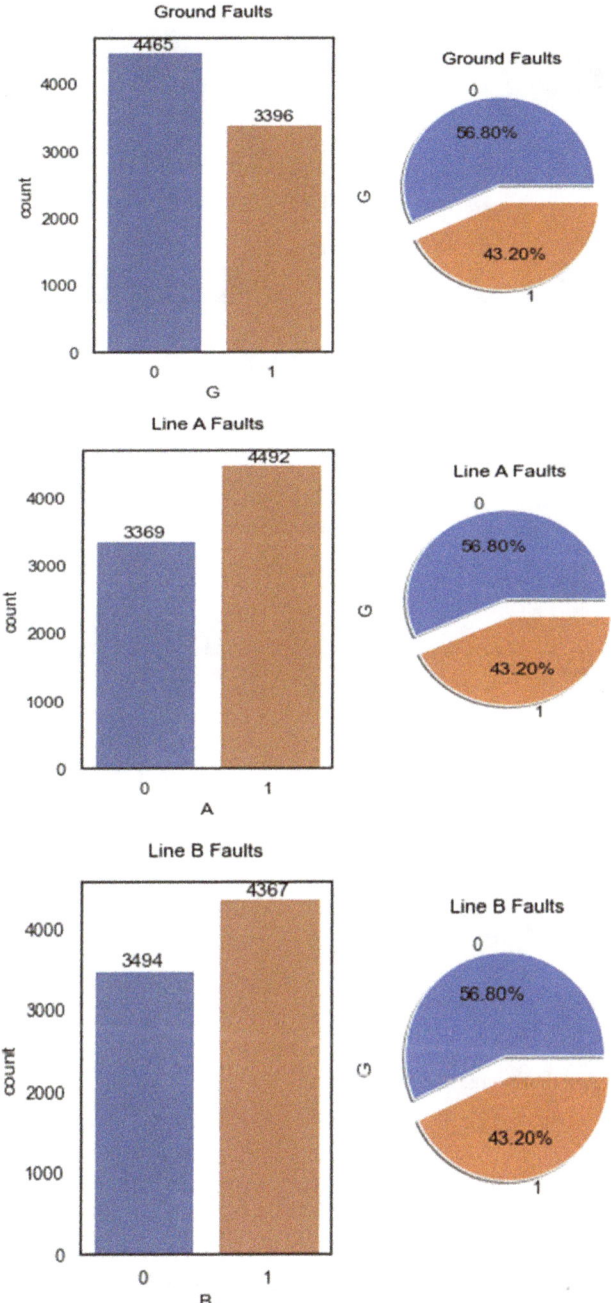

Figure 14.6. Data visualization of faulty transmission lines.

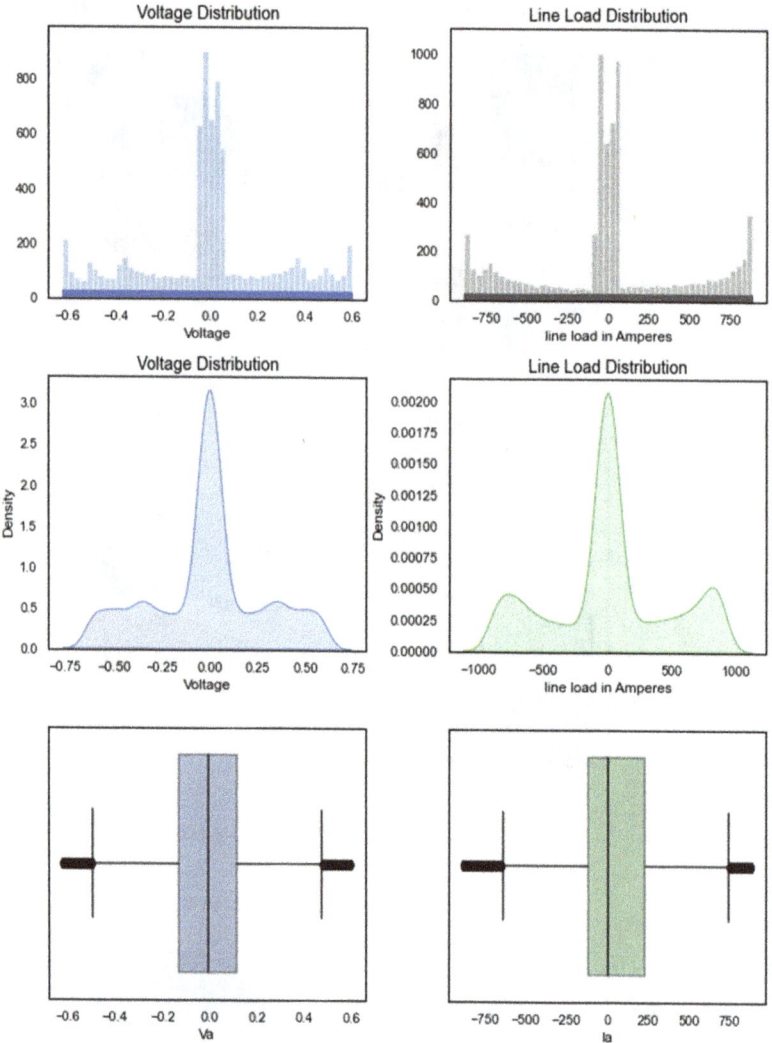

Figure 14.7. Visualizing the voltage distribution along with line load distribution in order to obtain results graphically.

Figure 14.9 presents the fault results after implementing the algorithm to obtain a complete picture of existing faults. 30.09% of lines have no faults. The maximum faulty system in lines A, B, and C to ground fault is 14.41%. Minimum faulty system is line B to line C fault 12.77%. The respective fault count in each transmission line is also shown with the help of a separate graph.

Table 14.1 presents the accuracies of different tested classification techniques which are obtained by a cross-validation method. It has been shown here through the help of ensembling what percentage of accuracy is incremented on a particular

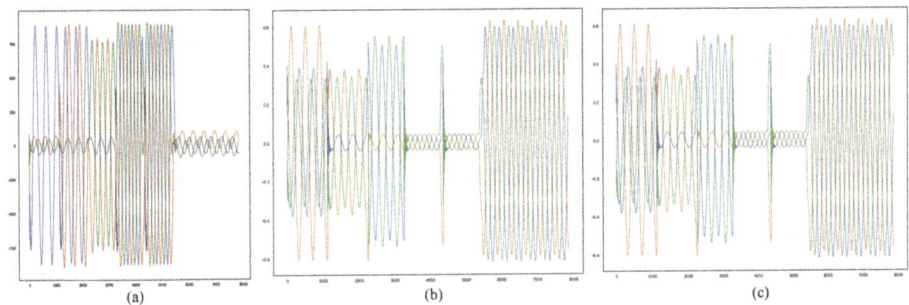

Figure 14.8. (a) Load distribution prevalent in power line A. The X-axis shows the time period. The Y-axis shows the amplitude. (b) Load distribution prevalent in power line B. The X-axis shows the time period. The Y-axis shows the amplitude. (c) Load distribution prevalent in power line C. The X-axis shows the time period. The Y-axis shows the amplitude.

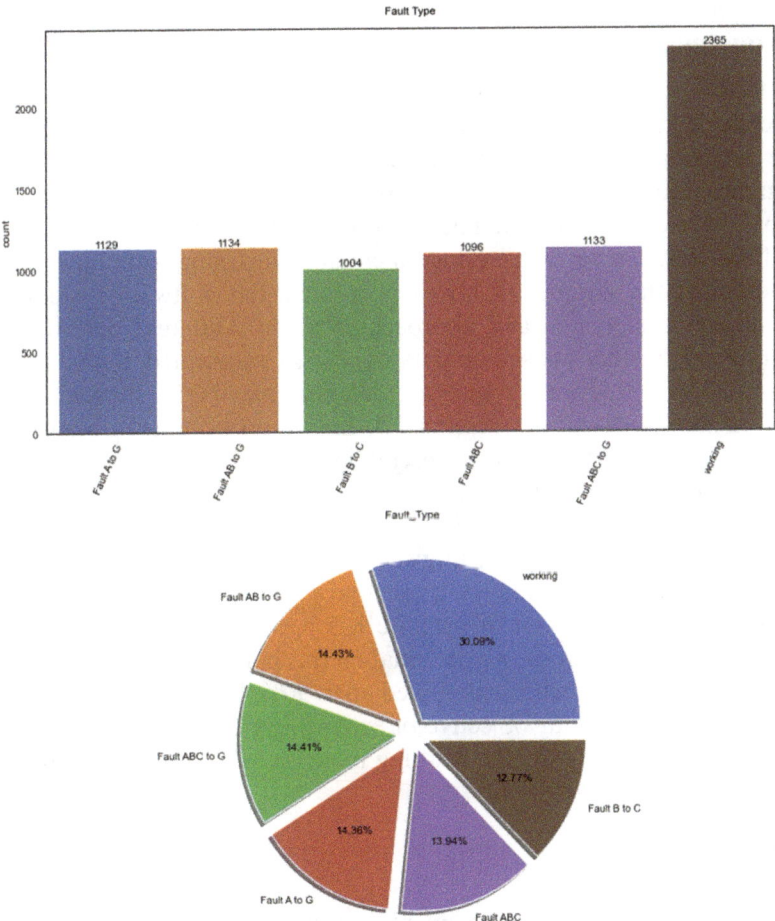

Figure 14.9. Fault results after the algorithm has been implemented are shown to obtain a complete picture of existing faults.

Table 14.1. Accuracy obtained using cross-validation.

Classification technique	Camel 1.3	Camel 1.4	Camel 1.5	Camel 1.6	Camel 1.7
Decision tree	75.55	71.15	71.51	76.11	78.61
Naïve Bayes	80.03	76.30	64.59	74.97	63.75
Logistic regression	76.76	72.75	73.79	77.94	73.56
Bagged-decision tree	83.58	85.37	79.31	84.99	86.41
Bagged-naïve Bayes	89.92	85.13	85.69	88.10	77.42
Bagged-logistic regression	84.30	80.39	78.73	85.08	87.06

prediction model of machine learning regressor [32]. In other words, ensemble learning models perform well, and they find and predict more problems thus far.

14.5 Conclusion

Fault detection is examined using two separate real-world scenarios. We have seen how machine learning is used in this defect detection and removal procedure. Some machine learning methods were discussed briefly. We observed how such fault detection technologies might be used to oversee electrical cables that typically have several problems during their lifetime. In addition, we observed how software might contain flaws that can be found using an ensemble learning technique. We may be able to estimate the number of errors in a particular software system by doing suitable dataset exploration and approach selection. Different detecting methods were considered. Methods based on data were less complicated. Signal models were more appropriate. Regressors such as decision tree (DT) showed the highest accuracy on the camel dataset version 1.7; the lowest accuracy obtained by them was 71%. It has been observed that ensembling proved beneficial.

The outcome can be seen in terms of increased accuracy of the same model when bagging is applied. A relevant increase of 14% can be seen in one of the cases where a DT regressor works on camel 1.4. Logistic regression did not perform much better, but when an ensemble was created, its accuracy was increased, i.e. from as low as 72% to as high as 87%. A lot of promise has been observed for the naïve Bayes regressor. This is because it shows potential on this particular dataset (camel) with a high accuracy of 80% and when ensembled up to 90%. For hardware fault detection, we successfully achieved our objective of exploring the power line dataset through visual exploration. Also, machine learning models could predict the faulty lines as they were intended to be able to.

Acknowledgments

I wish to acknowledge the encouragement from my organization for such a research opportunity.

References

[1] Zebin T, Scully P J and Ozanyan K B 2017 Inertial sensor based modelling of human activity classes: feature extraction and multi-sensor data fusion using machine learning algorithms *eHealth 360°* (Berlin: Springer) pp 306–14

[2] Zang Z, Li D and Wang J 2015 Learning classifier systems with memory condition to solve non-Markov problems *Soft Comput.* **19** 1679–99

[3] Zainudin Z and Shamsuddin S M 2016 Predictive analytics in Malaysian dengue data from 2010 until 2015 using BigML *Int. J. Adv. Soft Comput. Appl* **8** 19–30

[4] Wilson S W 1995 Classifier fitness based on accuracy *Evol. Comput.* **3** 149–75

[5] Urbanowicz R J and Moore J H 2009 Learning classifier systems: a complete introduction, review, and roadmap *J. Artif. Evol. Appl.* **2009** 1

[6] Tharwat A, Gaber T, Ibrahim A and Hassanien A E 2017 Linear discriminant analysis: a detailed tutorial *AI Commun.* **30** 169–90

[7] Stolzmann W 1998 Anticipatory classifier systems *Genetic Programm.* **98** 658–64

[8] Tabassum S, Sampa M B, Islam R, Yokota F, Nakashima N and Ahmed A 2020 A data enhancement approach to improve machine learning performance for predicting health status using remote healthcare data *2nd Int. Conf. on Advanced Information and Communication Technology (ICAICT)* pp 308–312

[9] Robertson G G and Riolo R L 1988 A tale of two classifier systems *Mach. Learn.* **3** 139–59

[10] Riolo R L 1991 Lookahead planning and latent learning in a classifier system *From Animals to Animats: Proceedings of the First International Conference on Simulation of Adaptive Behavior* pp 316–26

[11] Raschka S, Patterson J and Nolet C 2020 Machine Learning in Python: Main Developments and Technology Trends in Data Science, Machine Learning, and Artificial Intelligence *Information* **11** 193

[12] Noor N Q M, Sjarif N N A, Azmi N H F M, Daud S M and Kamardin K 2017 Hardware Trojan identification using machine learning-based classification *J. Telecommun., Electron. Comp. Eng. (JTEC)* **9** 23–7

[13] Navlani A 2018 Decision tree classification in Python

[14] Nagwanshi K K and Dubey S 2018 Statistical feature analysis of human footprint for personal identification using BigML and IBM Watson analytics *Arab. J. Sci. Eng.* **43** 2703–12

[15] Maleki M, Manshouri N and Kayikçioğlu T 2017 Application of PLSR with a comparison of MATLAB classification learner app in using BCI *25th Signal Processing and Communications Applications Conf. (SIU)*

[16] Liu W Z and White A P 1994 The importance of attribute selection measures in decision tree induction *Mach. Learn.* **15** 25–41

[17] Li M, Zhen L and Yao X 2017 How to read many-objective solution sets in parallel coordinates [educational forum] *IEEE Comput. Intell. Mag.* **12** 88–100

[18] Kessel M, Ruppel P and Gschwandtner F 2010 BIGML: a location model with individual waypoint graphs for indoor location-based services *PIK-Praxis der Informationsverarbeitung und Kommunikation* **33** 261–7

[19] Hussain A and Vatrapu R 2014 Social data analytics tool (sodato) *Int. Conf. on Design Science Research in Information Systems*

[20] Hoyt R E, Snider D H, Thompson C J and Mantravadi S 2016 IBM Watson analytics: automating visualization, descriptive, and predictive statistics *JMIR Public Health Surveill* **2** e157

[21] Holmes J H, Durbin D R and Winston F K 2000 The learning classifier system: an evolutionary computation approach to knowledge discovery in epidemiologic surveillance *Artif. Intell. Med.* **19** 53–74

[22] Holland J H *et al* 2000 What is a learning classifier system? *Learning Classifier Systems* (Berlin: Springer-Verlag)

[23] Holland J H, Holyoak K J, Nisbett R E and Thagard P R 1989 *Induction: Processes of Inference, Learning, and Discovery* (Cambridge, MA: MIT Press)

[24] Holland J H 1983 Escaping brittleness *Proc. 2nd Int. Workshop on Machine Learning*

[25] Holland J 1976 *Adaptation: Progress in Theoretical Biology* ed R Rosen and F M Snell (New York: Academic)

[26] Hallinan J S 2012 Data mining for microbiologists *Methods in Microbiology* (Amsterdam: Elsevier) vol 39 pp 27–79

[27] Goldberg D E and Holland J H 1988 Genetic algorithms and machine learning *Mach. Learn.* **3** 95–9

[28] Edsall R M 2003 The parallel coordinate plot in action: design and use for geographic visualization *Comput. Stat. Data Anal.* **43** 605–19

[29] De Mántaras R L 1991 A distance-based attribute selection measure for decision tree induction *Mach. Learn.* **6** 81–92

[30] Chappell D 2015 Introducing azure machine learning *A Guide for Technical Professionals, Sponsored by Microsoft Corporation* (Microsoft)

[31] Butz M V 2015 Learning classifier systems *Springer Handbook of Computational Intelligence* (Berlin: Springer) pp 961–81

[32] Acharya M S, Armaan A and Antony A S 2019 A comparison of regression models for prediction of graduate admissions *ICCIDS 2019: IEEE Int. Conf. on Computational Intelligence in Data Science* (Chennai, India)

Chapter 15

ACPSOD-Net: a deep atrous convolution pooling based network for salient object detection

Bhagyashree V Lad, Mohammad Farukh Hashmi and Avinash G Keskar

Systems incorporating a convolutional neural network (CNN) architecture are frequently utilized by many salient object detection (SOD) applications. When identifying the salient objects, detecting objects in diverse imaging scenarios is vital. To efficiently extract features and fuse them effectively, current SOD techniques generally use CNNs. Numerous strategies have been studies to further improve CNN representation in light of recent research that shows how well CNNs handle the edges and texture of images. This chapter introduces an original technique for identifying salient objects based on atrous convolution-based pooling and feature fusion networks to effectively integrate various features of objects. The proposed architecture is computationally very light and requires fewer learnable parameters. The chapter proposes a method based on an encoder–decoder network to detect salient objects. The encoders belong to the five ResNet-34 architecture's encoder blocks, and feature maps obtained by the encoders are passed through the atrous convolution-based pooling (ACP) block. The various features at the encoder step are integrated using a fusion network to obtain better saliency results. The fused feature maps are passed through five decoders to obtain the resultant saliency map. The outcomes of this study are assessed upon three well-known salient object detection datasets which illustrate the potential of the suggested system to accurately detect and localize the salient objects in different imaging scenarios.

15.1 Introduction

Salient object detection (SOD) seeks to spot the things in an image that are the most visually appealing. SOD is a crucial component of many computer vision and image processing operations as a pre-processing phase. Many applications, including image captioning [1], individual re-recognition [2], scene analysis and classification [3], object segmentation [4], visual tracking [5], and surface defect detection [6] make use of saliency as a pre-processing stage and concentrate on the input's greatest descriptive areas for accelerating the operation performance. The work on SOD is difficult since it involves the collection and integration of feature contents that comprise the local saliency information and the global image view.

The majority of traditional salient object recognition methods employ a bottom-up approach, classifying the input as relevant or not based on both cognitive and biological visual cues such as brightness, color, contrast, texture, etc. The traditional SOD methods are easy to use and consume fewer resources, but they are less accurate in some difficult circumstances since they have so little effective semantic insight. The primary way that traditional approaches detect salient objects is by employing hand-crafted features. Therefore, because high-level feature engineering is not present, their ability to characterize situations is somewhat limited, and their prediction performance is particularly weak in complex settings. Many of the techniques use machine learning based techniques [7, 8] along with traditional hand-crafted features to detect the salient objects which give good performance. In recent years fully convolutional networks (FCNs) broke the bottleneck of the old approaches and significantly aided in the development of SOD. FCNs [9] have the important potential to adaptively acquire excellent discriminative characteristics constituting an elaborate feature construction guided by annotated images. Recent studies have shown that end-to-end FCN-based techniques for SOD tasks outper-form conventional techniques and deep learning based on region supervision techniques in terms of detection accuracy. In several SOD algorithms, the signifi-cance of incorporating multiple-resolution contextual data was taken into account. As a way to extract complicated characteristics for minimizing crowded informa-tion, Wang *et al* [10] developed a contextual weighting block similar to that of Inception. The block used multi-branch convolutions with various kernel sizes to acquire receptive fields of various sizes. A bigger kernel size is required and important image content details are lost when all kernels are sampled at the same center. To collect multi-resolution salient objects, SOD models such as those in [11] and [12] incorporated the pyramid pooling module (PPM) [13]. The module combined various scenarios based on sub-regions. However, due to extensive pooling procedures, it loses the spatial features required for the SOD objective. The SOD work utilizes the atrous spatial pyramid pooling (ASPP) modules [14], which is also another well-liked multi-resolution contextual extraction tool. To adequately increase the receptive fields without incurring additional costs, the ASPP module used several concurrent atrous convolutions with various atrous rate values. The resulting context-pertinent characteristics were created by concatenating the outputs from these branches. To address the density and range problems associated with ASPP, Yang *et al* [15] densely coupled simultaneous atrous convolution operations. A simultaneous arrangement of two atrous convolutional layers with small but differing atrous rate values is made, *n*, [16] to create a layer of the improved ASPP. To create final representations of features, three identical layers with increasing dilation rates were closely coupled in succession.

Even though the FCN-based SOD approaches have demonstrated strong feature representation capabilities and made considerable advancements, identifying salient objects in challenging situations and low-contrast environments remains a challenge for SOD modeling. Also as discussed above, to incorporate the fine details, the majority of ASPP variations tried to increase the density of the sampling distribution. When performing conceptual segmentation tasks, in which a caption is assigned to each additional pixel within the image, such compact modules were demonstrated to be useful. Redundant information would arise if there were additional information flow channels used to consolidate metadata from a level into the entry for the following step. In this situation, a system must be created to limit undesired traits while encouraging advantageous ones for dynamic fusion with information from a subsequent higher scale. In this chapter, we have used atrous convolution-based pooling at each encoder stage and we have adaptively fused the features of these encoder stages to obtain the final saliency map. To obtain an accurate saliency map for various image conditions such as low-contrast images, complex patterned images, pictures with minute items that stand out, and pictures with many conspicuous objects, we have provided side supervision along with the supervision of the final saliency map.

The following is a summary of this work's key contributions:

- We demonstrate a novel network called a deep atrous convolution pooling based network for salient object detection (ACPSOD-Net) for SOD work. This design starts with the ResNet-34 [17] encoder architecture and makes use of a pooling block with atrous convolution [18] to widen the receptive field and capture more dependable and diversified data for better distinction.
- The proposed technique provides side output supervision which is summed to final output supervision to obtain better saliency results in different real-life imaging scenarios.
- Surpassing eight state-of-the-art benchmarking SOD techniques using three widely used, publicly accessible SOD datasets, ACPSOD-Net demonstrates exceptional performance.
- Extensive experimental test findings on three well-known SOD datasets demonstrated ACPSOD-Net's superior generalization capabilities. The suggested network model's emphasis on the target salient objects and their boundaries was demonstrated by exploring deep characteristics using qualitative performance, demonstrating the model's visual interpretability, and quantitative analysis, explaining the model's effectiveness in comparison to cutting-edge SOD techniques depending on various evaluation parameters used for the SOD task.

15.2 Proposed method

The proposed ACPSOD-Net's topology and main components are shown in figure 15.1. It uses a ResNet-34 [17] encoder and follows an encoder–decoder architecture where a decoder structure is explained further. An input image I of size $h \times w \times 3$ is first passed through the encoder architecture of ResNet-34 which

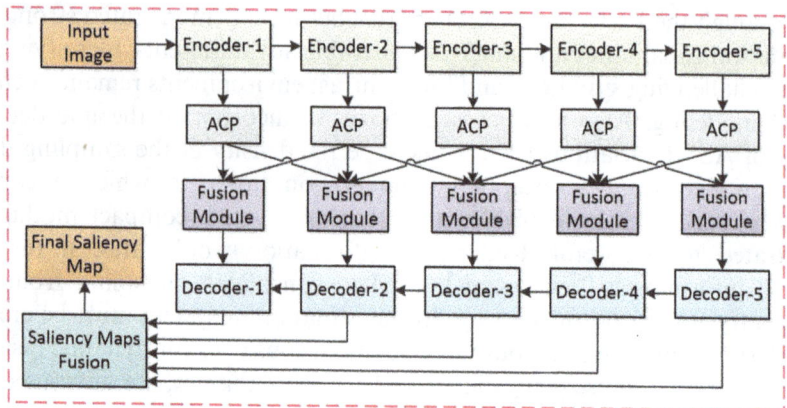

Figure 15.1. Workflow of the proposed ACPSOD-Net.

generates five consecutive feature maps. These feature maps are then passed through an atrous convolution-based pooling (ACP) module to obtain a wide range of multiple-scale features from the features of the encoders. The output features obtained from the ACP module are passed through the fusion module to obtain more descriptive global information-based fused feature maps. The final feature maps from these fusion modules are finally passed through the decoder block to obtain the required results. The results obtained by each decoder black are finally passed through the saliency map fusion module to obtain the resultant saliency map. The structure of each module, i.e. encoders, ACP, fusion module, decoders, and saliency map fusion, is explained further in figure 15.1.

15.2.1 Encoders

The conv1 and conv_2x layers of the ResNet-34 architecture make up the Encoder-1 where there are some changes in the conv1 layer of ResNet-34. Convolution in conv1 uses a 3×3 kernel with channels as 64 and stride as 1, and the max-pooling layer is removed rather than as used in ResNet-34, i.e. convolution with 7×7 kernel, channels as 64, and stride as 2, which is followed by max-pooling layer. The proposed architecture's Encoder-2, Encoder-3, and Encoder-4 are identical to the conv_3x, conv_4x, and conv_5x layers of ResNet-34 architecture. In a max-pooling layer, three fundamental residual blocks with 512 channels make up the last encoder, i.e. Encoder-5.

15.2.2 The atrous convolution-based pooling (ACP) module

The described technique's structure for ACP is depicted in figure 15.2. Five parallel 3×3 convolution layers with dilatation rates of 1, 3, 6, 9, and 12 make up the first part of the ACP module. The 3×3 kernel can cover a larger region over the entry-level features because of using atrous convolution which yields increased coverage of the feature space of the source. Due to this, by increasing the atrous rate, each layer's feature maps become better. Batch normalization and activation functions are used

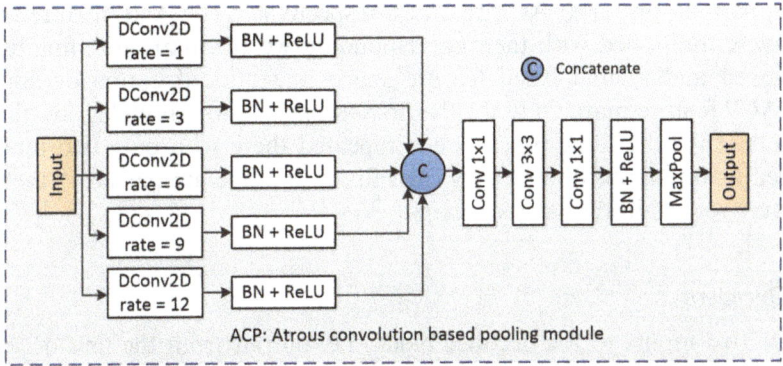

Figure 15.2. Architecture of atrous convolution-based pooling.

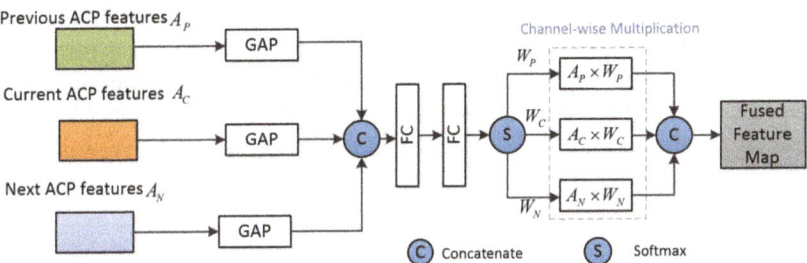

Figure 15.3. Architecture of the fusion module.

after every output of atrous convolution before being combined. The result obtained by each ReLU activation function are then combined using concatenation to create a single feature map, which is then processed via a series of 1×1, 3×3, and 1×1 convolutions to obtain elaborative combined features with a minimized number of feature channels. Batch normalization along with the ReLU activation function are added after the last 1×1 convolutional operation. To minimize the spatial dimensions of feature spaces, the outcome belonging to the ReLU activation function follows a max-pooling layer. This whole ACP architecture allows us to obtain a more informative multi-scale feature map corresponding to each encoder feature map.

15.2.3 Fusion module

The fusion module in the proposed work is the same as the adjacent layer attention block used in [19]. Here the current, previous, and next feature maps of the ACP modules are fused using some adaptive weights to obtain the final resultant fused feature map. The fusion framework's structure is demonstrated in figure 15.3. The fusion module generates the weight maps corresponding to each ACP feature, A_P, A_C, and A_N, as mentioned in figure 15.3. Here A_P, A_C, and A_N stand for the previous, current, and next ACP features, respectively. The ACP feature maps are channel-wise multiplied with their corresponding weight maps and, finally, all are concatenated and a final fused feature space is produced corresponding to the current ACP feature map. Only for the first and last fusion modules, are there only two inputs. Thus the same procedure is repeated there with only two inputs. The final fused feature maps obtained by each fusion module are passed to the decoder architecture to obtain the saliency maps.

15.2.4 Decoders

There are two inputs to the decoder block, i.e. output from the first decoder and output from the corresponding fusion module. Only Decoder-5 has single input, i.e. output from the corresponding fusion module. The decoder block first performs bilinear upsampling of an outcome of the earlier decoder module. An upsampled previous decoder outcome and fusion module outcome are then concatenated. A series of two residual blocks are applied to the concatenated feature maps extracting further significant semantic data contained in input and, finally, it is passed through the attention mechanism explained in [20]. To draw attention to the most important characteristics and to hide the unimportant ones, it first gives channel attention and then spatial attention. This decoder architecture generates five saliency maps concerning each decoder block.

15.2.5 Saliency map fusion

The five saliency maps obtained by corresponding decoder blocks are finally integrated to produce a resultant saliency map as a final output. This saliency map fusion module first upsamples and then concatenates all the saliency maps. The final saliency result is obtained by passing the discussed concatenated outcome through 3×3 convolution and the sigmoid activation function.

15.2.6 Loss function

The loss function is modified to be a mixed loss related to recent SOD research [21]. Binary cross-entropy (BCE) loss L_B and intersection over union L_I loss make up the modified mixed loss. The expression for the final loss function is

$$L = L_{B_{FS}} + L_{B_{SO}} + L_{I_{FS}}. \tag{15.1}$$

Here $L_{B_{FS}}$ stands for final saliency map binary cross-entropy loss, $L_{B_{SO}}$ stands for side output saliency map binary cross-entropy loss, and $L_{I_{FS}}$ stands for final saliency map intersection over union loss. To find the side output loss all the losses corresponding to each side output are added to obtain the resultant side output loss. The binary cross-entropy loss and intersection over union loss are calculated using expressions mentioned in the following two equations, respectively, where $Gt_{x,y}$ represents the ground-truth map and $Sa_{x,y}$ represents the saliency map:

$$L_B = -\sum_{x,y} Gt_{x,y} \log\left(Sa_{x,y}\right) + \left(1 - Gt_{x,y}\right) \log\left(1 - Sa_{x,y}\right) \qquad (15.2)$$

$$L_I = 1 - \frac{\sum\limits_{x,y} Gt_{x,y} Sa_{x,y}}{\sum\limits_{x,y} [Gt_{x,y} + Sa_{x,y} - Gt_{x,y} Sa_{x,y}]}. \qquad (15.3)$$

15.3 Experimental findings

Here we have compared the mentioned ACPSOD-Net with other SOD techniques already in use while conducting a quantitative and qualitative assessment. All investigation was carried out on a Windows-based PC with an NVIDIA GeForce RTX 3080Ti GPU and CUDA 10.2.

15.3.1 Implementation details and SOD datasets

According to the most popular salient object detection techniques being used currently, we employ as training images the 10 553 images of the DUTS-TR [22] dataset. The ECSSD [23], DUT-OMRON [24], and HKU-IS [25] datasets, each containing 1000, 5168, and 4447 images, respectively, were used to test the ACPSOD-Net's performance. We employed the F-measure (F_β), mean absolute error (MAE), enhanced alignment matrix measure, i.e. E-measure (Em) [26], and structural measure (Sm) [27] scores as the assessment metrics to evaluate the objective analysis. The PyTorch framework is utilized for the training of the ACPSOD-Net. As described in [28], we employed intersection over union loss and binary cross-entropy loss (BCE loss with side output deep supervision) as loss functions to optimize the stochastic gradient descent (SGD) optimization algorithm. The batch size is fixed to 8, the learning rate is selected as 10^{-4}, and the highest epoch number is adjusted to 100.

15.3.2 Results and discussion

We have subjectively and objectively evaluated the proposed ACPSOD-Net against eight cutting-edge deep learning based SOD approaches, which are by Feng *et al* [29], Zhao *et al* [30], Qin *et al* [31], Zhao *et al* [32], Chen *et al* [33], Pang *et al* [34], Liu *et al* [35], and Wu *et al* [36]. Table 15.1 shows the objective analysis based on the MAE and F_β scores, whereas table 15.2 shows the objective analysis based on Sm and Em scores. According to the objective results, the proposed ACPSOD-Net outperforms or performs on par with several current cutting-edge SOD approaches, including Liu *et al* [35] and Wu *et al* [36]. Several SOD methods, including Feng *et al* [29], Qin *et al* [31], and Zhao *et al* [30] operate on edge-aware SOD challenges in which they primarily require additional flow for recognizing edge cues. The ACPSOD-Net method improves the F_β score by $\approx 2.11\%$–5.11%, the MAE score by $\approx 13.9\%$–26.36%, the Sm score by $\approx 1.07\%$–2.36%, and the Em score by

Table 15.1. Objective analysis of the proposed ACPSOD-Net using MAE and F_β score. (Red, green, and blue are used to rank the model at the top three positions. A lower MAE score and higher F-measure score are preferable.)

Datasets	DUT-OMRON		HKU-IS		ECSSD	
Evaluation metrics	F_β	MAE	F_β	MAE	F_β	MAE
Feng et al [29]	0.717	0.057	0.869	0.036	0.886	0.042
Zhao et al [30]	0.738	0.053	0.887	0.031	0.903	0.037
Qin et al [31]	0.751	0.056	0.889	0.032	0.904	0.037
Zhao et al [32]	0.729	0.055	0.88	0.033	0.894	0.04
Chen et al [33]	0.745	0.058	0.883	0.036	0.882	0.042
Pang et al [34]	0.738	0.056	0.897	0.029	0.911	0.033
Liu et al [35]	0.755	0.058	0.897	0.029	0.91	0.033
Wu et al [36]	0.77	0.050	0.908	0.027	0.918	0.033
ACPSOD-Net	0.768	0.054	0.912	0.024	0.915	0.034

Table 15.2. Objective analysis of the proposed ACPSOD-Net using Sm and Em scores. (Red, green, and blue are used to rank the model at the top three positions. Higher Sm and Em scores are desirable.)

Datasets	DUT-OMRON		HKU-IS		ECSSD	
Evaluation metrics	Sm	Em	Sm	Em	Sm	Em
Feng et al [29]	0.826	0.846	0.905	0.935	0.913	0.935
Zhao et al [30]	0.841	0.857	0.918	0.944	0.925	0.943
Qin et al [31]	0.836	0.865	0.909	0.943	0.916	0.943
Zhao et al [32]	0.838	0.855	0.915	0.937	0.92	0.936
Chen et al [33]	0.825	—	0.908	—	0.917	—
Pang et al [34]	0.833	0.86	0.919	0.952	0.925	0.95
Liu et al [35]	0.85	0.871	0.928	0.952	0.932	0.951
Wu et al [36]	0.849	0.878	0.924	0.955	0.927	0.951
ACPSOD-Net	0.851	0.881	0.929	0.953	0.926	0.944

$\approx$ 0.97%–2.1% when compared to these edge-aware SOD methodologies. When compared to previous cutting-edge SOD approaches, the research framework also avoids using extra deep flow to detect edges, which minimizes the architectural complexity and several trainable parameters, simultaneously increasing the salient object recognition precision by ACPSOD-Net. The subjective outcomes of ACPSOD-Net over cutting-edge SOD methods are shown in figure 15.4. In this chapter, we have shown three instances of salient object recognition in four different contexts: an image with a complicated background, an image with very little object-to-background contrast, an image containing multiple objects, as well as an image

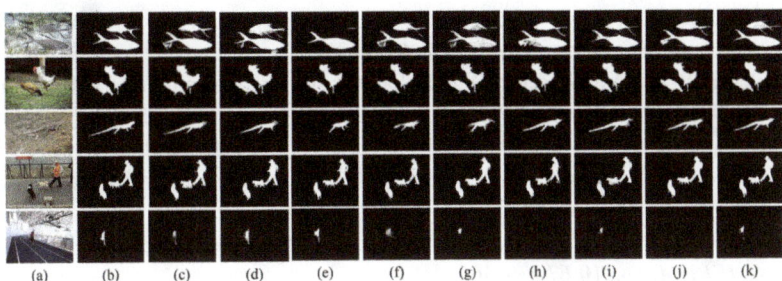

Figure 15.4. Subjective results of ACPSOD-Net compared to cutting-edge SOD techniques: (a) original image, (b) ground-truth, (c) Feng *et al* [29], (d) Zhao *et al* [30], (e) Qin *et al* [31], (f) Zhao *et al* [32], (g) Chen *et al* [33], (h) Pang *et al* [34], (i) Liu *et al* [35], (j) Wu *et al* [36], and (k) ACPSOD-Net.

containing a salient object that is extremely small. In all four instances, as is evident from figure 15.4, the suggested technique recognizes salient objects containing the right discrimination between foreground and background. Similar findings are obtained using the Wu *et al* [36] and Liu *et al* [35] approaches; however, as shown in figure 15.4, they are incapable of recognizing very small salient objects. Therefore, it can be concluded from the subjective and objective evaluation that the proposed ACPSOD-Net offers improved SOD findings with better retention of the objects' details.

15.4 Conclusion

In this research we investigated SOD tasks on real-world images. The proposed ACPSOD-Net is made up of encoders built on the ResNet-34 architecture, an ACP to increase the receptive field of the encoder features, a fusion module to adaptively fuse the adjacent features to obtain more information, decoders to generate the corresponding saliency maps, and the fusion block to combine all the output results generated by decoders. To create accurate and detail-retaining saliency results, the proposed ACPSOD-Net is helpful. The proposed ACPSOD-Net's superiority and effectiveness in accurately detecting the salient items in contrast to several cutting-edge SOD approaches are demonstrated by extensive subjective and objective experimentation on three difficult SOD datasets.

References

[1] Xu J, Ba R, Kiros K, Cho A, Courville R, Salakhudinov R, Zemel Y and Bengio 2015 Show, attend and tell: neural image caption generation with visual attention *Proc. Mach. Learn. Res.* **37** 2048–57

[2] Zhao R, Ouyang W and Wang X 2013 Unsupervised salience learning for person re-identification *Proc. IEEE Conf. on Computer Vision and Pattern Recognition* (Piscataway, NJ: IEEE) pp 3586–93

[3] Ren Z, Gao S, Chia L T and Tsang I W H 2013 Region-based saliency detection and its application in object recognition *IEEE Trans. Circuits Syst. Video Technol.* **24** 769–79

[4] Wang W, Shen J, Yang R and Porikli F 2017 Saliency-aware video object segmentation *IEEE Trans. Pattern Anal. Mach. Intell.* **40** 20–33

[5] Mahadevan V and Vasconcelos N 2009 Saliency-based discriminant tracking *2009 IEEE Conf. on Computer Vision and Pattern Recognition* (Piscataway, NJ: IEEE) pp 1007–13

[6] Song G, Song K and Yan Y 2020 EDRNet: encoder–decoder residual network for salient object detection of strip steel surface defects *IEEE Trans. Instrum. Meas.* **69** 9709–19

[7] Lad B V, Hashmi M F and Keskar A G 2022 Boundary preserved salient object detection using guided filter based hybridization approach of transformation and spatial domain analysis *IEEE Access* **10** 67 230–46

[8] Lad B V, Das M, Hashmi M F, Keskar A G and Gupta D 2022 Saliency detection using a bio-inspired spiking neural network driven by local and global saliency *Appl. Artif. Intell.* **36** 2094408

[9] Long J, Shelhamer E and Darrell T 2015 Fully convolutional networks for semantic segmentation *Proc. IEEE Conf. on Computer Vision and Pattern Recognition* (Piscataway, NJ: IEEE) pp 3431–40

[10] Wang T, Zhang L, Wang S, Lu H, Yang G, Ruan X and Borji A 2018 Detect globally, refine locally: a novel approach to saliency detection *Proc. IEEE Conf. on Computer Vision and Pattern Recognition* (Piscataway, NJ: IEEE) pp 3127–35

[11] Liu J J, Hou Q, Cheng M M, Feng J and Jiang J 2019 A simple pooling-based design for real-time salient object detection *Proc. IEEE/CVF Conf. on Computer Vision and Pattern Recognition* (Piscataway, NJ: IEEE) pp 3917–26

[12] Zhang L, Wu J, Wang T, Borji A, Wei G and Lu H 2020 A multistage refinement network for salient object detection *IEEE Trans. Image Process.* **29** 3534–45

[13] Zhao H, Shi J, Qi X, Wang X and Jia J 2017 Pyramid scene parsing network *Proc. IEEE Conf. on Computer Vision and Pattern Recognition* (Piscataway, NJ: IEEE) pp 2881–90

[14] Chen L C, Papandreou G, Kokkinos I, Murphy K and Yuille A L 2017 DeepLab: semantic image segmentation with deep convolutional nets, atrous convolution, and fully connected CRFS *IEEE Trans. Pattern Anal. Mach. Intell.* **40** 834–48

[15] Yang M, Yu K, Zhang C, Li Z and Yang K 2018 DenseASPP for semantic segmentation in street scenes *Proc. IEEE Conf. on Computer Vision and Pattern Recognition* (Piscataway, NJ: IEEE) pp 3684–92

[16] Wang L, Chen R, Zhu L, Xie H and Li X 2020 Deep sub-region network for salient object detection *IEEE Trans. Circuits Syst. Video Technol.* **31** 728–41

[17] He K, Zhang X, Ren S and Sun J 2016 Deep residual learning for image recognition *Proc. IEEE Conf. on Computer Vision and Pattern Recognition* (Piscataway, NJ: IEEE) pp 770–8

[18] Zhang J, Lin S, Ding L and Bruzzone L 2020 Multi-scale context aggregation for semantic segmentation of remote sensing images *Remote Sens.* **12** 701

[19] Gupta A K, Seal A, Khanna P, Herrera-Viedma E and Krejcar O 2021 Almnet: adjacent layer driven multiscale features for salient object detection *IEEE Trans. Instrum. Meas.* **70** 1–14

[20] Woo S, Park J, Lee J-Y and Kweon I S 2018 CBAM: convolutional block attention module *Proc. European Conf. on Computer Vision (ECCV)* pp 3–19

[21] Wei J, Wang S and Huang Q 2020 F3Net: fusion, feedback and focus for salient object detection *Proc. of the AAAI Conf. on Artificial Intelligence* vol 34 pp 12 321–8

[22] Wang L, Lu H, Wang Y, Feng M, Wang D, Yin B and Ruan X 2017 Learning to detect salient objects with image-level supervision *Proc. IEEE Conf. on Computer Vision and Pattern Recognition* (Piscataway, NJ: IEEE) pp 136–45

[23] Shi J, Yan Q, Xu L and Jia J 2015 Hierarchical image saliency detection on extended CSSD *IEEE Trans. Pattern Anal. Mach. Intell.* **38** 717–29

[24] Yang C, Zhang L, Lu H, Ruan X and Yang M H 2013 Saliency detection via graph-based manifold ranking *Proc. IEEE Conf. on Computer Vision and Pattern Recognition* (Piscataway, NJ: IEEE) pp 3166–73

[25] Li G and Yu Y 2015 Visual saliency based on multiscale deep features *Proc. IEEE Conf. on Computer Vision and Pattern Recognition* (Piscataway, NJ: IEEE) pp 5455–63

[26] Fan D-P, Gong C, Cao Y, Ren B, Cheng M M and Borji A 2018 Enhanced-alignment measure for binary foreground map evaluation arXiv:1805.10421

[27] Fan D-P, Cheng M M, Liu Y, Li T and Borji A 2017 Structure-measure: a new way to evaluate foreground maps *Proc. IEEE Int. Conf. on Computer Vision* (Piscataway, NJ: IEEE) pp 4548–57

[28] Shen K, Zhou X, Wan B, Shi R and Zhang J 2022 Fully squeezed multiscale inference network for fast and accurate saliency detection in optical remote-sensing images *IEEE Geosci. Remote Sens. Lett.* **19** 1–5

[29] Feng M, Lu H and Ding E 2019 Attentive feedback network for boundary-aware salient object detection *Proc. IEEE/CVF Conf. on Computer Vision and Pattern Recognition* (Piscataway, NJ: IEEE) pp 1623–32

[30] Zhao J X, Liu J J, Fan D P, Cao Y, Yang J and Cheng M M 2019 EGNet: edge guidance network for salient object detection *Proc. IEEE/CVF Int. Conf. on Computer Vision* (Piscataway, NJ: IEEE) pp 8779–88

[31] Qin X, Zhang Z, Huang C, Gao C, Dehghan M and Jagersand M 2019 BASNet: boundary-aware salient object detection *Proc. IEEE/CVF Conf. on Computer Vision and Pattern Recognition* (Piscataway, NJ: IEEE) pp 7479–89

[32] Zhao X, Pang Y, Zhang L, Lu H and Zhang L 2020 Suppress and balance: a simple gated network for salient object detection *European Conf. on Computer Vision* (Berlin: Springer) pp 35–51

[33] Chen S, Tan X, Wang B, Lu H, Hu X and Fu Y 2020 Reverse attention-based residual network for salient object detection *IEEE Trans. Image Process.* **29** 3763–76

[34] Pang Y, Zhao X, Zhang L and Lu H 2020 Multi-scale interactive network for salient object detection *Proc. IEEE/CVF Conf. on Computer Vision and Pattern Recognition* (Piscataway, NJ: IEEE) pp 9413–22

[35] Liu N, Zhang N, Wan K, Shao L and Han J 2021 Visual saliency transformer *Proc. IEEE/CVF Int. Conf. on Computer Vision* (Piscataway, NJ: IEEE) pp 4722–32

[36] Wu Y H, Liu Y, Zhang L, Cheng M M and Ren B 2022 EDN: salient object detection via an extremely-downsampled network *IEEE Trans. Image Process.* **31** 3125–36

Printed in the USA
CPSIA information can be obtained
at www.ICGtesting.com
CBHW082046010424
5693CB00045B/275